W0234736

What Every Engineer Should Know About Developing Plastics Products

WHAT EVERY ENGINEER SHOULD KNOW
A Series

Editor

William H. Middendorf

Department of Electrical and Computer Engineering
University of Cincinnati
Cincinnati, Ohio

Vol. 1   What Every Engineer Should Know About Patents, *William G. Konold, Bruce Tittel, Donald F. Frei, and David S. Stallard*

Vol. 2   What Every Engineer Should Know About Product Liability, *James F. Thorpe and William H. Middendorf*

Vol. 3   What Every Engineer Should Know About Microcomputers: Hardware/ Software Design: A Step-by-Step Example, *William S. Bennett and Carl F. Evert, Jr.*

Vol. 4   What Every Engineer Should Know About Economic Decision Analysis, *Dean S. Shupe*

Vol. 5   What Every Engineer Should Know About Human Resources Management, *Desmond D. Martin and Richard L. Shell*

Vol. 6   What Every Engineer Should Know About Manufacturing Cost Estimating, *Eric M. Malstrom*

Vol. 7   What Every Engineer Should Know About Inventing, *William H. Middendorf*

Vol. 8   What Every Engineer Should Know About Technology Transfer and Innovation, *Louis N. Mogavero and Robert S. Shane*

Vol. 9   What Every Engineer Should Know About Project Management, *Arnold M. Ruskin and W. Eugene Estes*

Vol. 10 What Every Engineer Should Know About Computer-Aided Design and Computer-Aided Manufacturing: The CAD/CAM Revolution, *John K. Krouse*

Vol. 11 What Every Engineer Should Know About Robots, *Maurice I. Zeldman*

Vol. 12 What Every Engineer Should Know About Microcomputer Systems Design and Debugging, *Bill Wray and Bill Crawford*

Vol. 13 What Every Engineer Should Know About Engineering Information Resources, *Margaret T. Schenk and James K. Webster*

Vol. 14 What Every Engineer Should Know About Microcomputer Program Design, *Keith R. Wehmeyer*

Vol. 15 What Every Engineer Should Know About Computer Modeling and Simulation, *Don M. Ingels*

Vol. 16 What Every Engineer Should Know About Engineering Workstations, *Justin E. Harlow III*

Vol. 17 What Every Engineer Should Know About Practical CAD/CAM Applications, *John Stark*

Vol. 18 What Every Engineer Should Know About Threaded Fasteners: Materials and Design, *Alexander Blake*

Vol. 19 What Every Engineer Should Know About Data Communications, *Carl Stephen Clifton*

Vol. 20 What Every Engineer Should Know About Material and Component Failure, Failure Analysis, and Litigation, *Lawrence E. Murr*

Vol. 21 What Every Engineer Should Know About Corrosion, *Philip Schweitzer*

Vol. 22 What Every Engineer Should Know About Lasers, *D. C. Winburn*

Vol. 23 What Every Engineer Should Know About Finite Element Analysis, *edited by John R. Brauer*

Vol. 24 What Every Engineer Should Know About Patents, Second Edition, *William G. Konold, Bruce Tittel, Donald F. Frei, and David S. Stallard*

Vol. 25 What Every Engineer Should Know About Electronic Communications Systems, *L. R. McKay*

Vol. 26 What Every Engineer Should Know About Quality Control, *Thomas Pyzdek*

Vol. 27 What Every Engineer Should Know About Microcomputers: Hardware/Software Design: A Step-by-Step Example. Second Edition, Revised and Expanded, *William S. Bennett, Carl F. Evert, and Leslie C. Lander*

Vol. 28 What Every Engineer Should Know About Ceramics, *Solomon Musikant*

Vol. 29 What Every Engineer Should Know About Developing Plastics Products, *Bruce C. Wendle*

Additional volumes in preparation

What Every Engineer Should Know About Developing Plastics Products

Bruce C. Wendle
Boeing Commercial Airplane
Seattle, Washington

Marcel Dekker, Inc.   New York • Basel • Hong Kong

Library of Congress Cataloging-in-Publication Data

Wendle, Bruce C.
    What every engineer should know about developing plastics products
    Bruce Wendle.
    6 p.   cm. — (What every engineer should know)
    Includes bibliographical references and index.
    ISBN 0-8247-8485-5
    I. Title. II. Series.
    TA455.P5W48   1991   91-20567
    668.4—dc20   CIP

This book is printed on acid-free paper.

Copyright © 1991 by MARCEL DEKKER, INC. All Rights Reserved

Neither this book nor any part may be reproduced or transmitted in any form
or by any means, electronic or mechanical, including photocopying, micro-
filming, and recording, or by any information storage and retrieval system,
without permission in writing from the publisher.

MARCEL DEKKER, INC.
270 Madison Avenue, New York, New York 10016

Current printing (last digit):
10 9 8 7 6 5 4 3 2 1

PRINTED IN THE UNITED STATES OF AMERICA

Preface

What every engineer should know about developing plastics products—an interesting topic when you consider how much of today's technology revolves around the various polymers that make up today's plastics. Hardly any product in use today is made without some form of synthetic material. From the highly specialized medical field, to transportation, to the construction industry, plastics have found uses, and those uses are continuing to grow.

It is impossible to write a book defining what one should know about an industry that is changing drastically and dramatically almost every day. Not only are new polymers being introduced continually, but the combination of various resins to form alloys and the output of resins with various additives, such as glass and carbon fibers, makes it nearly impossible for engineers to keep up.

You must then combine this new polymer technology with the introduction of the computer into the many divisions of the plastics industry. Add software packages that simulate the filling of a part on a computer, as well as techniques for designing parts and tooling, and you have a rapidly changing industry. The processing area is in a state of flux as well. New ways of forming a plastic part are being developed, and new auxiliary equipment is being combined with computer controlled molding equipment to change the face of plastics processing.

Some of the old tried-and-true methods are still in use and there is still no substitute for good old-fashioned experience. However, if you do not have a knowledge of these technologies, you are going to be left behind. That is why this book is a mixture of the old and the new. Its purpose is to bring engineers up to date with regard to existing technology while introducing them to innovations that are right around the corner. We have not only included the physical properties of many of the new polymers but have also covered methods of purchasing tooling, secondary techniques, and many ideas on how to work with design engineers to produce plastic products.

One cannot expect everything to go smoothly; unexpected events are bound to take place in the development process. Murphy's law operates in the plastics industry as it does in any other. What is supposed to happen often does not—you can count on it. This book will help you to be prepared for the unexpected while introducing you to the plastics industry of the future.

Bruce C. Wendle

Contents

Preface   *iii*

1. Plastics as a Material   1
2. Development of Plastic Products   15
3. Material Selection   39
4. Modifying Plastic   65
5. Plastic Processing   75
6. Secondary Operations   103
7. Tooling for Plastics   117
8. Dealing with the Plastics Industry   133

Appendixes
   I.   Selected Conversion Factors   161
   II. History of Plastics   163
   III. Selected Plastics Suppliers   171
   IV. Standards and Practices of Plastics Molders   173

Bibliography   *179*
Index   *183*

1

Plastics as a Material

PLASTICS IN OUR DAILY LIVES

No other material has played as important a part in our daily lives as have the synthetic polymers known as *plastics*. Metals, rock, and wood have all been around for a long time and each has brought us a myriad of usable products. But no material has so captured the minds of designers and engineers as has plastic. Over a few short years the material has gone from a laboratory curiosity to a preferred material of construction in almost every industry in the world. New uses are being found almost daily and whole new industries are being developed around plastic.

Used in the form of fiber, plastics have found their way into all kinds of textitles for both consumer and industrial uses. In extruded sheet or profile form, the material is used as a building material, pipe, conduit, glazing, and for a variety of other uses. In molded form the number of applications are too numerous to mention. It is safe to say that there is virtually no modern industry that has not considered the use of plastics in some form or another.

Recent developments in the plastics field have been numerous and more are coming out every day. For example, in the area of composites, where new materials such as carbon glass and fibers have been combined with such well-known polymers as epoxy, polyesters, urethanes, and even

Figure 1
The introduction of the computer into the plastics industry has changed the way parts are designed, the way tools are designed, the way plastic parts are processed, and the way parts are checked for proper dimensions. (Photo courtesy of SPI.)

some thermoplastics, the combinations are endless. These materials make it possible to develop many new applications never before thought possible. Aircraft components and automotive engine parts are just a few areas now opening up for plastics. As plastics approach the strength of metals, while providing lighter weights and easier production methods, many new and exciting application areas will open up. To be a part of this important revolution is an exciting and fascinating experience. Today's plastic engineers not only have to know their plastics technology but must be conversant in many other technologies as well.

Figure 2
The wide use of plastics on this tractor exemplifies the potential for engineering resins in the design of lawn and gardening equipment. Designed for maximum function and efficiency, assembly has been reduced by 47%, and the part count has been reduced from 224 to 129. (Photo courtesy of General Electric.)

The use of new metals and other materials in tooling, of adhesives for bonding, and the introduction of new polymers all open up a need for developing new technologies to be combined with older methods. Use of the computer to design and control processing of plastic parts has brought about its own revolution. Tool design and processing simulation by computer have created new needs for qualified personnel not even imagined just a decade ago. As this new technology grows and matures, the need for plastic engineers trained in these new methods will grow.

DEFINITIONS

One can get many answers to the question: What are plastics? One dictionary defines plastics as "a synthetically produced material that can be molded and hardened into objects or formed into films or textile fibers." They are produced primarily from gas and oil products by large chemical companies throughout the world. Technically, the question is best answered by quoting Ronald D. Beck from his book *Plastic Product Design*: "Plastics are organic materials made from large molecules that are constructed by a chain-like attachment of certain building block molecules. The properties of the plastic depend heavily on the size of the molecule and on the arrangement of the atoms within the molecule" [11]. We know plastics as a modern-day material, capable of a wide range of properties and able to be formed or molded into a variety of shapes and sizes. Plastics touch all our lives in one manner or other.

As more synthetic polymers are developed, they tend to replace many of our more conventional materials, such as metal, glass, and paper. Combinations of polymers alloyed with carbon, glass, or other materials are definitely challenging metals under such severe conditions as the inside of a combustion engine or the structural member of an aircraft. It is becoming increasingly difficult for engineers to keep up with the latest developments in plastic technology. It is for this reason that this book was written.

Thermoplastics and Thermosets.   There are two basic types of plastics available from which to produce products. These are defined as thermoplastic and thermoset. *Thermoplastic* resins can be passed through a heat cycle and when cooled down, can be ground up and reused. *Thermosets*, on the other hand, can be passed through only one heat cycle; they cannot be re-formed. Over the years these two basic types of polymers have merged so that today many thermosets can be processed like thermoplastics, so there is no longer a clear distinction between the two categories. This is also true in the urethane area: Most urethanes are still considered a thermoset castable material, but some, including the injection-moldable elastomers, can be processed like a thermoplastic on an injection molding machine or extruder.

Each polymer has its own distinct characteristics and set of physical properties. The practice of alloying various polymers has opened up an entirely new range of materials. With the addition of such additives as carbon, glass, and metal fibers, the range of available properties of today's plastics has changed dramatically. The result of this is that today one can specify a given set of physical properties and nearly always find a polymer or alloy that will meet or surpass them. This has opened up many new application areas for plastics. The future seem bright for these new synthetic materials that only 120 years ago were available only for use in billiard balls.

Crystalline and Amorphous Polymers. In some thermoplastics, the chemical structure is such that the polymer chains will fold in on themselves and pack together in an organized manner. The resulting organized regions show the behavioral characteristics of crystals. Plastics that have these regions are called *crystalline* polymers; plastics without them are called *amorphous*. All crystalline plastics have amorphous regions between and connecting the crystalline regions. For this reason the crystalline polymers are often referred to as *semicrystalline*.

Liquid Polymers. Liquid crystalline polymers are best thought of as being a separate and unique class of plastic. The molecules are stiff, rodlike structures that are organized in large parallel arrays of domains in both the melted and solid states. This makes possible a unique material with high values of several physical properties. Readers can obtain more information on liquid crystal polymers from Dartco Manufacturing, Celanese, or Eastman Chemical (see also Ref. 2).

MECHANICAL AND PHYSICAL PROPERTIES

Many of the mechanical and physical property differences between plastics can be attributed to their structure. As a generalization, the ordering of crystalline and liquid crystalline thermoplastics makes them stiffer, stronger, and less impact resistant than their amorphous counterparts. Also, crystalline and liquid crystalline polymers have a higher resistance to creep, heat, and chemical exposure. However, crystalline materials are more difficult to process because they have higher melt temperatures and tend to shrink and warp more than do amorphous polymers.

NEW DEVELOPMENTS IN POLYMERS

In recent years the demand for higher-heat-resistant materials with no loss in physical properties has pushed the development of new polymers and alloys. In the commercial aircraft industry, this demand for improvements has brought about the development of such materials as polyether ether ketone (PEEK), polyether imide (PEI), and polyamide imide (PAI). These new materials have heat-deflection temperatures of up to 600 °F. They also have improved chemical resistance and maintain high physical property values. Versions are available with glass and carbon to provide increased rigidity and improved dimensional stability.

However, these high-heat-resistant engineering materials have changed the normal processing conditions found with conventional plastics. Equipment with molding temperatures of 1000 °F have had to be developed and mold temperature controllers have had to be designed to operate in the 300 °F range. (An English–metric conversion table is provided as Appendix I.) This has been made easier by the advent of computer-controlled molding equipment.

Uses for these materials include computer and business machine housings and components, aircraft parts and automotive under-the-hood applications. Included also are industrial and electrical applications that require added heat resistance. Special grades of these new materials are available that meet the demanding standards set by the Federal Aviation Agency. Underwriters' Laboratories, and other controlling agencies. Included among these are stringent flammability requirements that require plastics to be made less flammable and able to withstand higher heats. The molder should be checked for the availability of the necessary equipment and experience to run these materials before molded or extruded parts made of these new polymers are requested.

HISTORY OF PLASTICS

To better understand the development of the plastics industry, one should be aware of the development of various applications and their relationship to the introduction of new polymers. A compilation of the highlights of developments in the plastics industry over the last 125 years is included in Appendix II. A quick glance through these exciting developments provides a better understanding of the history of plastics and provides a perspective on where new applications might fit into its future.

Significant Developments in the Plastics Industry. Since the early twentieth century, when cellulosics were developed to produce billiard balls, the plastic industry has been formed and shaped by one exciting development after the other. First came the development of materials to produce a growing number of applications, then the equipment necessary to form and mold these new polymers. These were the events that shaped today's plastic industry. Significant developments along the way included the introduction of the in-line screw machine, which allowed molders to control melt conditions more carefully while processing the materials. Then came the computer, which allowed molders a great deal more control over processing and provided the capabiity of monitoring the materials on a minute-by-minute basis. It also allowed moldmakers new opportunities for machining molds. These new developments are already having a far-reaching effect on the industry—and there is still more to come.

Today's Plastic Industry. Today's plastic industry is made up of several distinct layers, each with a special service to perform. The first layer comprises the material suppliers, usually large chemical companies with millions of dollars invested in chemical plants and distributing systems spread throughout the world. These companies, including General Electric and Du Pont, are dedicated to supplying the basic raw materials and resins that form the foundation of the industry. They also supply the basic polymer research, new application development, and design assistance needed by their customers to utilize successfully the materials produced.

The second layer consists of material distributors and reprocessors. These companies distribute smaller quantities of the resins produced by the major chemical companies and process and sell a variety of combinations of these materials. They make and sell glass-, metal-, and carbon-filled varieties of these products together with flame-retarded grades and special lubricated polymers. They generally are also the source of small lots of custom-colored resins. As the costs of producing and selling basic resins have skyrocketed, the large chemical companies have relied more and more on distributors and reprocessors to

Figure 3
The newest music device, the compact disk, is now being manufactured from
an engineering thermoplastic. (Photo courtesy of Mobay Chemical.)

get their materials to market. Working hand in hand with various color
houses, these distributors have become a virtual grocery store of plastics,
making available everything from smaller lots of basic resins to
precolored and color concentrates for all kinds of specialized products.

The processors are divided into two groups: the large, captive molders
and extruders and the smaller, more numerous custom molders. The
captive shops are usually owned by large manufacturing firms and tend
to concentrate on one or more product lines. The custom shops specialize
in providing products for a large variety of industries and will usually
manufacture whatever is needed to satisfy their customers.

Supplying both these types of firms are a variety of moldmaking shops, equipment manufacturers, and specialized supply houses, all providing a string of products, tooling, and equipment needed by the molding houses. A number of the molding companies cross over and produce their own tooling. Large national manufacturing companies such as automotive firms have set up their own molding facilities to provide captive product lines. Others include the beverage bottle manufacturers and manufacturers of household goods, such as Rubbermaid.

Engineers seeking employment can find anything they want in the various layers of the plastic industry. From the small five- to six-machine custom molder to the largest chemical companies, all offer exciting opportunities for engineers and designers interested in plastics.

ENVIRONMENTAL STATUS OF PLASTICS

The use of plastics affects the environment in many ways. In many cases, plastics are good for the environment. Trash bags, filters, and containers of all kinds are important in controlling waste management. Also plastics will not rust or oxidize, so they can often be used where metals would not be considered.

The flammability of plastics plays an important part in the environment as well as affecting the general safety of products in which it is used. As an example, the new FAA regulations requiring all commercial aircraft to meet new high flammability and heat load test standards, as well as restricting the amount of smoke and toxicity, will play an important part in making our travel environment safer. Other restrictions have been set as to which plastics can be used safely by such organizations as Underwriters' Laboratories.

All plastic products burn. This is not surprising when one considers that most are produced from petroleum and natural-gas raw materials. The main product of combustion when plastic is burned in an excess of oxygen or air is carbon dioxide. Other more toxic products can result when plastics are burned in an oxygen-restricted atmosphere. It is therefore important to know what the fire scenario is before talking about specific poisonous gases being given off by the burning of plastic materials. Various materials can be added to plastic to reduce burning but may cause the emission of poisonous gases and often reduce the physical properties.

Waste disposal management is another area where plastics play an important role in our environment. For years plastics have been going into landfills. The basic polymers generally do not break down, and some say that this is bad for the environment. Materials have been developed that mix starch molecules with the plastic molecules so that when exposed to starch-eating bacteria, they are broken into small particles.

Synthetic polymers do break down when exposed to the ultraviolet radiation present in our sunlight. Parts with molded-in or application-induced stress are very likely to fail and crack when exposed to sunlight. The addition of inhibitors and other methods of protection, such as the addition of carbon black, often reduces stress-induced failures. Painting a plastic product will also protect it from degradation.

Recycling of plastics is also becoming a popular pro-environmental action. Companies exist that separate polymers from waste and by one means or another clean and chop the waste plastics into a form that can be sold to molders for reprocessing into useful products. Three basic polymers—polyethylene, polyester, and polyvinyl chloride—are presently available in this form and others will be separated soon.

You can see that many aspects of plastics affect our enviornment. Often one action will trigger other more detrimental results. It is therefore important to consider the whole picture before deciding what effect plastics are having on our environment.

THE FUTURE

Recent developments in the computer-aided design and manufacturing field will bring about significant changes in the way in which plastic parts are designed, prototyped, and built. One of the first developments in this area was the introduction of Moldflow, a software package developed in Australia and offered to the plastics industry as a way to predict the way a specific material would flow into a part. Using this computer simulation, a design may make changes in the design of a part before metal is actually cut for the tooling.

Over the last few years, other software packages have been developed that carry the concept much further. Unisys, Inc. has come out with a system that not only analyzes material flow, but provides a way to design the mold and analyze the cooling system, and instructs a

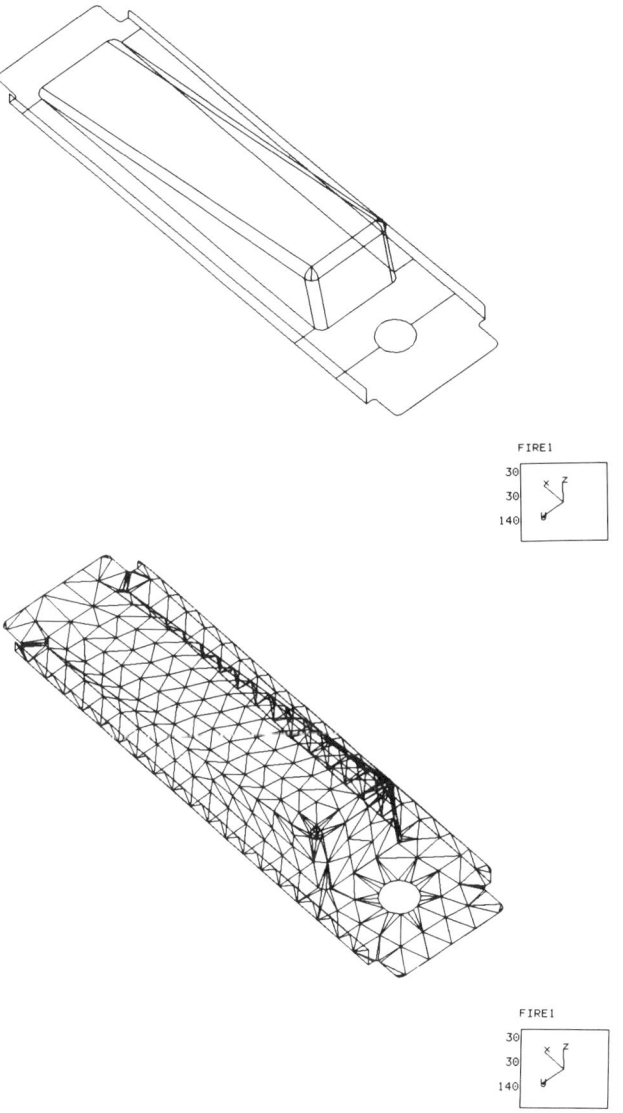

Figure 4

Using a CAD system, the object to be designed is first built as a wire frame
model (top). The model is divided into finite elements (bottom) by using a
software package such as Moldflow. The computer is then used to determine
such values of temperature time, or pressure at each finite element. Thus the
complete mold filling can be simulated before any metal tooling is built.

numerically controlled (NC) mill as to how to machine the cavity and core. Such systems are rather expensive, but like other programs of this type, are getting less expensive as they become better known. Competition in the computer software industry is very strong, and this will bring down the cost. Whether these systems will eventually replace the experienced hands-on plastic engineer remains to be seen, but it will certainly bring about changes in the way that parts and tooling are designed and constructed.

From an advertising brochure published by Unisys comes the following: "With our plastics solution [new software package] you can produce parts more quickly while reducing costs because it lets you graphically design, draft, and analyze parts and molds quickly and easily without even building a physical prototype.... Exclusive plastics programs for mold filling and cooling analysis assure superior mold performance prior to production, and a numerical control package speeds higher precision mold machining" [3].

Another company that presents a complete evaluation for a plastic part is Technical Molding Concepts. Their software program also assists in the development of a part design. It evaluates the material requirements for a product and suggests the best material to use. It then assists you by simulating the filling of a mold, gives you tooling design assistance, and analyzes the cooling system with the purpose of developing the most economical design.

Another significant change that is taking place is the combination of what were formerly separate job functions within the plastics industry. With the advent of the computer, it is becoming more important to combine the part design function with production and tooling design jobs. It is not enough just to be familiar with part design; you must understand processing of the polymer as well. You should also have a good grasp of tooling and tooling design. The computer will help you do this job, but you must understand the technology and be able to use the computer for what it is—a tool.

The Plastics Plant of the Future.   It is important here to stop and try to visualize just what is in store for the plastics industry in the future. Certainly there will be many changes. The high cost of equipment will almost certainly bring an end to the many small two- and three-machine job shops that today make up a large share of the plastics industry.

Development of a molding system that will introduce a polymer in liquid form, which will be piped into an extruder-like reactor and polymerized just before introduction into the mold is a probability. Called *in situ polymerization*, the system is like the reaction injection molding (RIM) process already available and used extensively by the automotive industry. By changing the stream of additives and varying the polymer ratio, the properties of the output polymer can be changed. Products from solid to foam would be possible, all controlled and monitored by computer. The computer monitoring system would tell management exactly what equipment is running, how well, and keep track of jobs throughout the plant. All shipping and receiving, as well as billing and accounts payable, would be done by computer.

Automatic mold changing systems, now available, would be utilized extensively and setup time would be kept to a minimum. Robots would be used in place of operators on all machines and would be used a great deal for secondary operations as well. In the field of moldmaking, most molders would build their own tooling, Computer-aided design and manufacturing would be the rule rather than the exception, and most shops would be able to go from their computer to the NC mill to produce tooling. Lead time to build molds would thus be cut substantially.

All of these changes are available in some form right now. The trick will be to get them all working in an organized manner. The opportunities for designers and engineers of all types will be great, and plastic applications will be many and varied.

REFERENCES

1. Ronald Beck. *Plastic Product Design*, 2nd ed., Van Nostrand Reinhold Co., Inc., New York, 1980.
2. *Designing with Plastic: The Fundamentals*, Engineering Plastics Division. Hoechst Celanese, 1988.
3. *Why Complete CAD/CAM Solutions Are More Sensible*, Unisys, 2970 Wilderness Place, Boulder, CO 80301.

2

Development of Plastic Products

In this chapter we define what constitutes a plastic application and some of the problems involved in developing and producing a plastic part. We have selected an application produced by injection molding because it is the most widely used form of plastic processing. Applications using other processes, such as extrusion, blow molding, and vacuum forming, would be developed in a similar manner.

THE BEGINNING

A project always starts with an idea: maybe something you have been thinking about for years, or perhaps an idea that came to you while you were in the shower. An idea often starts as a sketch on the back of a napkin or as a model of something you built to solve a problem. Whatever the case, if it can be built out of plastic, if it can be sold, and if it is something on which you can make money, it has possibilities.

DEFINITION OF A PROJECT

As mentioned earlier, a good plastics project is one that starts as an idea: either an improvement on something already in existence or, more

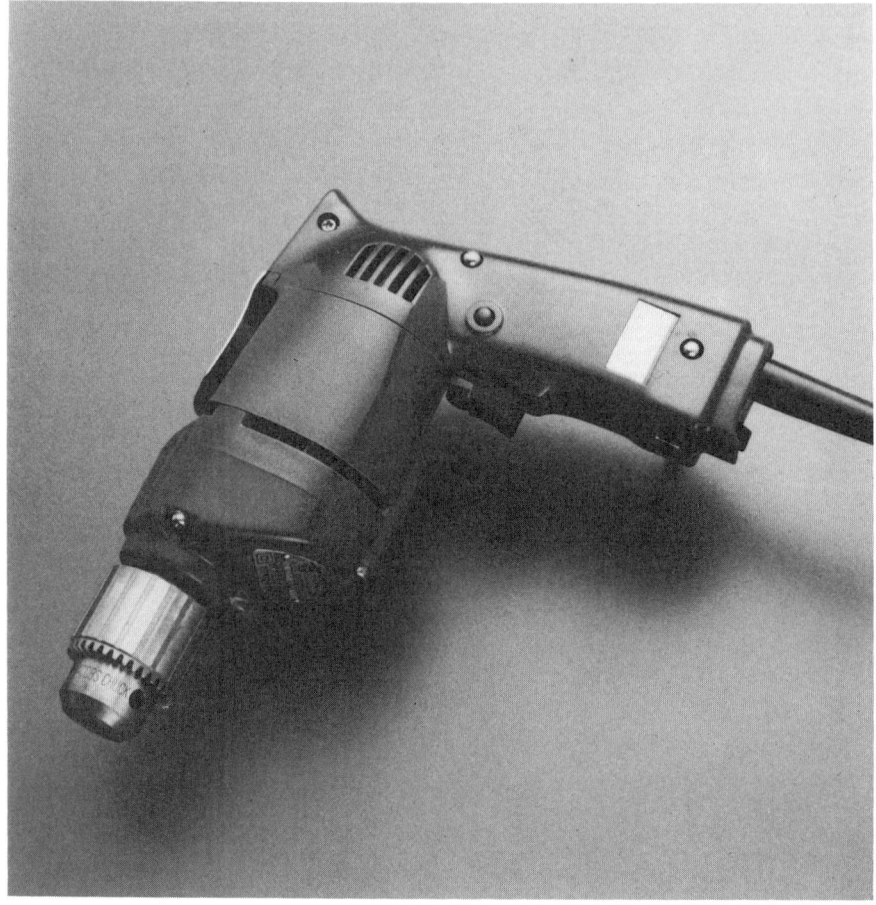

Figure 5

A ½-in. drill housing made of engineering material and designed to take tough use and provide an insulation at the same time. (Photo courtesy of Mobay Chemical.)

rarely, a truly unique way of accomplishing something new and worthwhile. From this beginning the idea is developed into a workable product. At this point there is obviously much to be done before you can think of the idea as an actual product.

Once the idea has been put into feasible, working form, it is time to determine just how the article is to be built and approximately how much it will cost to build it. A number of methods are open to the engineer to determine what will work. These include building a working model. This might not be built out of the material from which you will finally build the production item, but it should give you enough of an idea of the end product so that you will be in a position to show potential vendors what you have in mind.

Sometimes there still is not enough information to make a determination as to how the product should be built. Often, several ideas have to be investigated and more than one model built before a clear model can be formulated. The important thing is to get as many workable ideas as possible from people who are experts in their field.

You may already have decided how to build the product—but do not be satisfied with your own ideas. Make a list of all the good things about the idea and all the things that are wrong with it. Get as many opinions on the subject as you can. Often the "right" way will be obvious, but sometimes there will be several ways, all with good possibilities for success. Never overlook the basics, such as manufacturing cost, shipping costs, weight, and overall size. Any one of these items overlooked now could cause problems later. Get as much information as you can before you make a final decision. Keep an open mind on the various aspects of the project and do not be afraid to change something if a better way to do it shows up. Remember that almost everything has been tried before and there is no reason to reinvent the wheel.

EXAMPLE OF A PROJECT

Since we are considering only plastics at this time, we will stick to the development of such a product. One of the best ways to understand the process of developing a plastic project is to study an actual example of such a development.

A gentleman came into our molding facility with a basic idea for a new type of fishing lure designed to be used with live bait and aimed at attracting the large salmon so popular in the Puget Sound country of western Washington state. He had a working model of the lure machined from a block of clear solid plastic. Using this model, he had caught several large fish, and he was interested in producing the product in large quantities. We examined the lure and decided that the best way to produce the item was to develop a steel injection tool and to mold the lure out of a clear, strong plastic material.

First, we needed to develop a set of working drawings. This was an absolute necessity because without them neither the customer nor the molder would be sure of what was required. Before the drawings could be completed, the customer had to be sure of what he was going to have produced and the moldmaker had to know just what he was suppose to build. In the case of the fishing lure, this involved several more models being hand-machined out of plastic rod stock, and the added time required to test the new design in fishing waters.

Once the general configuration was established, the work of the designer began. The customer chose a product designer who had a computer-aided design (CAD) package on his computer. It was felt that a computer-aided design was needed to develop the complex curves and shapes present in the application. A contract was written to cover the design of the product.

Three or four weeks were allowed for the designer to complete work on the drawings. Once done, they were discussed thoroughly with the customer and then signed off. In this way it was established that the customer knew exactly what the molder was being asked to produce. The drawings were then given to the molders who had been selected, requesting a final quote on the tooling and a price to produce certain quantities of the lure. Also included in the quote was the cost of providing hardware and special jigging tools needed to finish the product.

Special steel staples had to be embedded in the plastic lure to hold the hook assembly. Also needed was a source for stainless steel threaded inserts and bolts to fasten the lure together once it was molded. There was some disagreement as to who should be responsible for these items, the molder or the customer. Some customers like to have an overall line of responsibility but want the molder to handle all details. Others like to be involved with the purchase of all items.

BRAINSTORMING

Often, it is a good idea to introduce an idea to a circle of friends or business associates so that the project can be considered from many viewpoints. You should gather these people together around a table away from the phone so that they can concentrate completely on your project. Be sure to write down every thought or idea that comes from the discussion and do not throw out any ideas that may appear silly or unworkable. Make sure that everyone participating understands the basics of the project fully, and then let the ideas flow.

Some questions that you could throw out to get the session started would be:

1. What does the product do?
2. Does it mate with other parts that could also be included?
3. What advantage would switching to plastic have on the application?
4. What are several ways in which the application could be produced?

I'm sure you get the idea. Just cover whatever comes into anybody's head and let the conversation roam from the obvious to the ridiculous. Keep all of the ideas and review them several times. A thought or comment will often trigger a solution long after the brainstorming session is over.

FIELD TESTING

Do not overlook the value of field testing a product. If practical, have a number of prospective customers try out your model. Remember to keep a good written record of all aspects of the test. This type of information is invaluable and makes critical decisions easier when the time comes.

INDUSTRIAL DESIGNERS

If your firm is not large enough to have your own design group, you may want to consider working with an outside industrial designer. Generally, you will want to choose one that has a good reputation and some experience in the process or material you are going to be using for your application. In today's world, most good designers use computer-aided design programs on their computers, and you may want to consider this in your decision.

Initially, the designer is going to want to know everything you can communicate about the application. Knowing as much as possible about the product's use will help in designing the part and in searching out aspects of the product that you might not have considered. Such other information as dimensional tolerances, environmental conditions to be experienced by the product, and volumes required will all need to be discussed.

The ouptut of the designer should be a set of engineering drawings defining the product in detail and giving information on the material from which the end product will be built. By this time you should have available sufficient information to go out and obtain a valid quotation. The design criteria and how to choose the right material and process are discussed later in the chapter.

THE QUOTATION

Once you have a complete set of engineering drawings and other information pertaining to your application, you are ready to obtain cost figures from vendors. You should pick prospective vendors carefully. Develop a list of these vendors you wish to have quote on the project and provide them all with the identical information. You should give the vendors sufficient time to develop their quotations and do not try to insist on getting the numbers in an unrealistic amount of time.

Be prepared to listen to vendors' ideas and to make changes in the product if requested. Often, a vendor has had considerable experience at producing the same type of product and can be of significant assistance in making the product less expensive and better. Once you have received the vendors' quotations, you are in a position to make a decision on the product. Since the molder is often the prime contractor on an application of this type, you will receive a quotation on the tooling as well as the cost of the product itself.

MECHANICAL AND PHYSICAL PROPERTIES

We cover the mechanical and physical properties of plastic in Chapter 3. These properties, which are usually determined by various mechanical tests, are very important to the performance of any application. Although the plastics industry does not have the years of experience and data that are available from the metals industry, there is still a substantial amount of good information available. Organizations such as the ASTM (American Society for Testing and Materials) have developed testing methods and published standards on all tests available on plastics.

As more and more has become known about plastics, design engineers and product designers have had to rely less and less on just plain luck and experience and have been able to utilize good physical data and accurate load and stress calculations to predict the performance of plastics. Although it is not within the scope of this book to get into the subject too deeply, one should know that this information exists. Several material suppliers have developed assistance in this area, and such services as General Electric's computer data bank are of real value when trying to get an understanding of what is available in this field.

DESIGN CONSIDERATIONS*

In the design or analysis of any mechanical component, a systematic approach is desirable. Frequently, the product will be one in which there are no significant loads and no deflection limitations. In these cases the "gut feeling" of the design engineer is often all that is required. This is especially true with small load-free parts where processing requirements dictate a minimum wall thickness which is more than adequate for the part function. Still, even in these cases, engineers new to plastic design often neglect the effects of stresses caused by assembly, handling, shipping, temperature changes, and other environmental changes. As the complexity of the part increases or when very accurate results are required, more exact classical methods of design are required.

Loads. The first step in analyzing any application is to determine the load that will be applied to the part. Directly applied loads are known loads applied to a defined area of the part, concentrated at a point, line, or boundary, or distributed over an area:

1. Both ends fixed and a concentrated load at the center (Fig. 7)
2. Both ends fixed and a uniformly distributed load (Fig. 8)
3. Simple supported beam and a concentrated load to the center (Fig. 9)
4. Cantilevered beam (one fixed end) with concentrated load at the free end (Fig. 10)

---

*Much of this information on design comes from an excellent design manual published by the Engineering Plastics Division of Hoechst Celanese. Entitled *Designing with Plastic: The Fundamentals*, the manual gives an excellent starting point for any engineer wishing to learn more about plastic design.

Figure 6
This model of a conceptual portable environmental control system developed
by GE Plastics demonstrates the benefits of engineering resins in designing
HVAC products for consumer appeal, as well as assembly and manufacturing
efficiency. The system shows how materials can reduce overall part count and
assembly time to lower costs. (Photo courtesy of General Electric.)

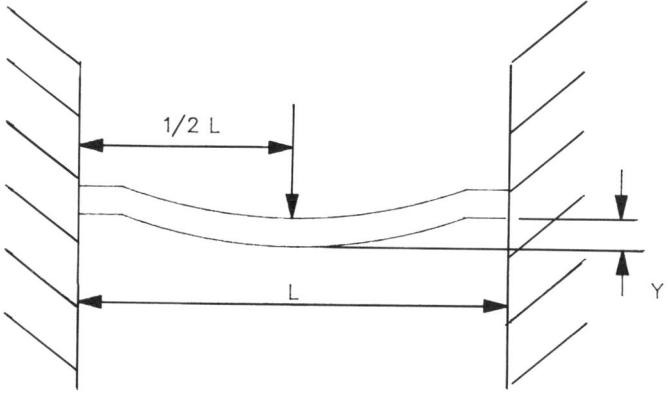

(at supports) $a = \dfrac{F\,L}{8\,Z}$

(at load) $Y = \dfrac{F\,L^3}{192\,E\,I}$

Figure 7
Both ends fixed and a concentrated load at the center.

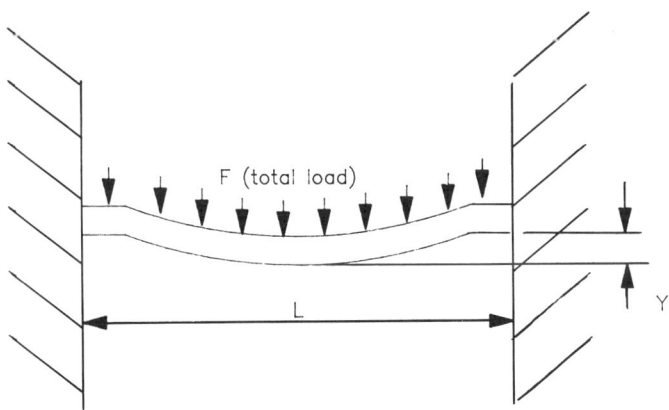

(at supports) $a - \dfrac{F\,L}{12\,Z}$

(at load) $Y = \dfrac{F\,L^3}{384\,E\,I}$

Figure 8
Both ends fixed and a uniformly distributed load.

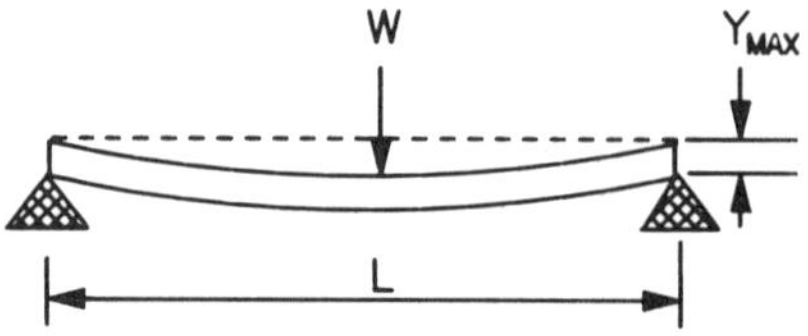

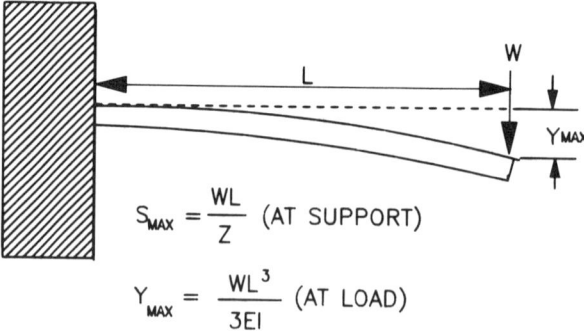

$$S_{MAX} = \frac{WL}{4Z} \quad \text{(AT LOAD)}$$

$$Y_{MAX} = \frac{WL}{48EI} \quad \text{(AT LOAD)}$$

Figure 9
Simple supported beam and a concentrated load to the center.

5. Simple supported beam with a uniformly distributed load (Fig. 11)
6. Cantilevered beam (one end fixed) with a uniformly distributed load
    (Fig. 12)

Pressure Vessels.   The most common plastics pressure vessel application is a tube with internal pressure. In selecting the wall thickness for the tube, the thin-walled tube hoop stress equation is convenient. It is very useful in determining an approximate wall thickness even when the wall thickness is less than the diameter divided by 10. After the thin-walled tube equations have been applied, the thick-walled tube equation can be used to verify the design.

$$S_{MAX} = \frac{WL}{Z} \quad \text{(AT SUPPORT)}$$

$$Y_{MAX} = \frac{WL^3}{3EI} \quad \text{(AT LOAD)}$$

Figure 10
Cantilevered beam (one fixed end) with a concentrated load at each end.

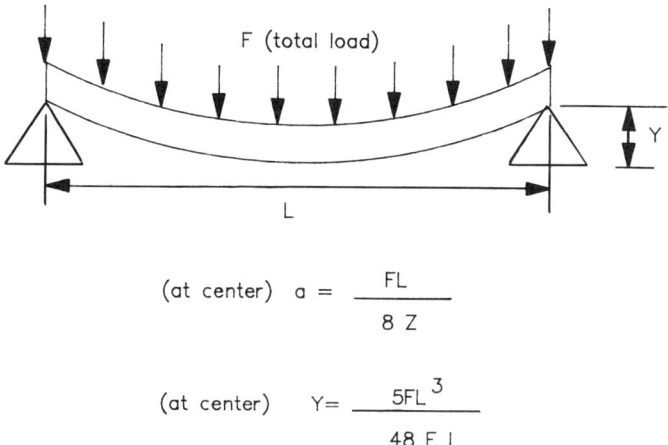

$$\text{(at center)} \quad a = \frac{FL}{8\,Z}$$

$$\text{(at center)} \quad Y = \frac{5FL^3}{48\,E\,I}$$

Figure 11
Simple supported beam with a uniformly distributed load.

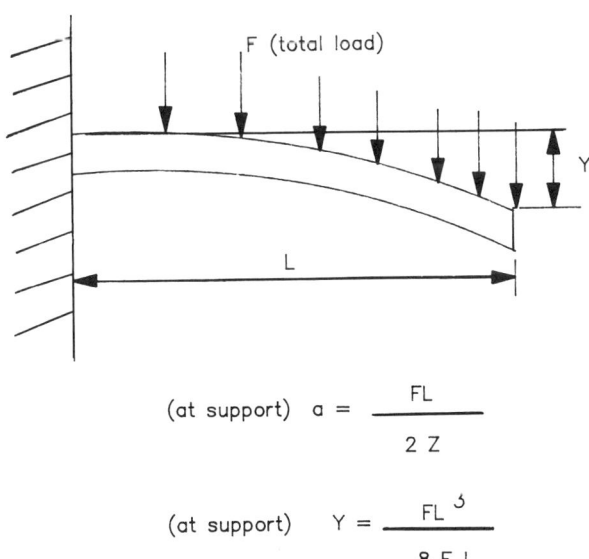

$$\text{(at support)} \quad a = \frac{FL}{2\,Z}$$

$$\text{(at support)} \quad Y = \frac{FL^3}{8\,E\,I}$$

Figure 12
Cantilevered beam (one end fixed) with a uniformly distributed load.

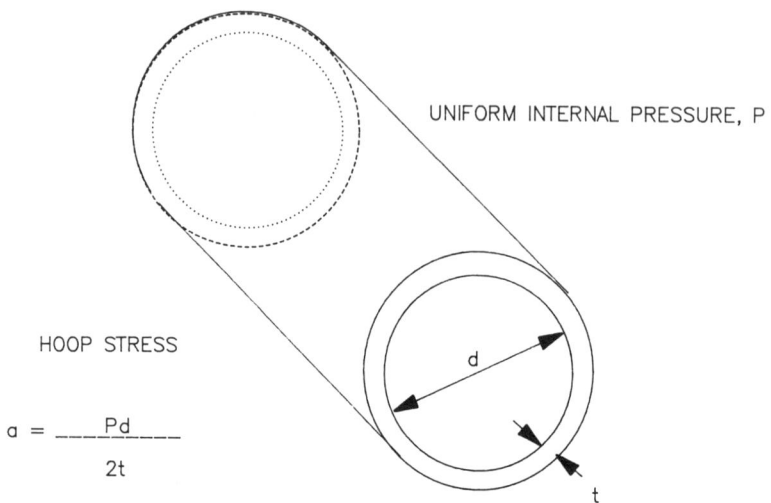

UNIFORM INTERNAL PRESSURE, P

HOOP STRESS

$$\sigma = \frac{Pd}{2t}$$

This equation is reasonably accurate for t>d/10. As the wall thickness increases the error becomes quite large.

Figure 13
Cylindrical pressure vessel with thin-walled tube.

A cylindrical pressure vessel with thin-walled tube is shown in Fig. 13. Fig. 14 shows a cylindrical pressure vessel with thick-walled tube.

Strain-Induced Loads. Frequently, a part becomes loaded when subjected to a defined deflection. The actual load is a result of the structural reaction of the part to the strain applied. Many assemby stresses and thermal stresses are a result of strain-induced loads.

Safety Factors. There are no hard and fast rules to follow in setting safety factors for plastic parts. Before putting any product into the marketplace, tests should be run on the actual parts as the most extreme operating conditions and in the presence of any chemicals that might be expected during the end use. Safety factors* for preliminary design work are as follows:

---

*Suggested percentage of strength values published in marketing data sheets are based on type of stress and maximum temperatures. These are intended for preliminary design analysis only and are not to be used in place of thorough product testing.

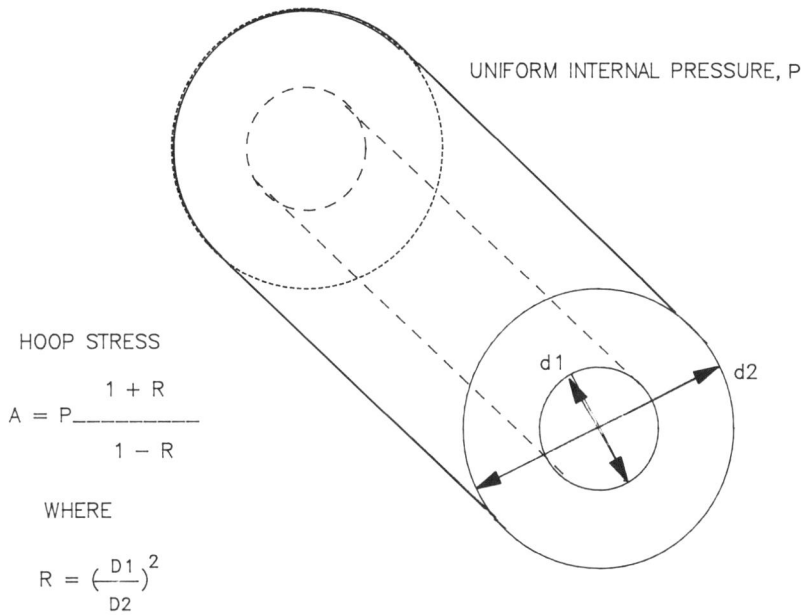

HOOP STRESS

$$A = P\frac{1 + R}{1 - R}$$

WHERE

$$R = (\frac{D1}{D2})^2$$

This equation is for the maximum hoop stress which occurs on the surface of the inside wall of the tube.

Figure 14
Cylindrical pressure vessel with thick-walled tube.

	Failure not critical (%)	Failure critical (%)
Intermittent loading (nonfatigue)	25–50	10–25
Continuous loading	10–25	5–10

Fig. 15 shows maximum stress and deflection equations for selected beams.

Draft. The draft on vertical walls should be considered carefully. A designer must make sure that there is enough draft to get the part from the mold. In most cases the molder is going to want as much draft as

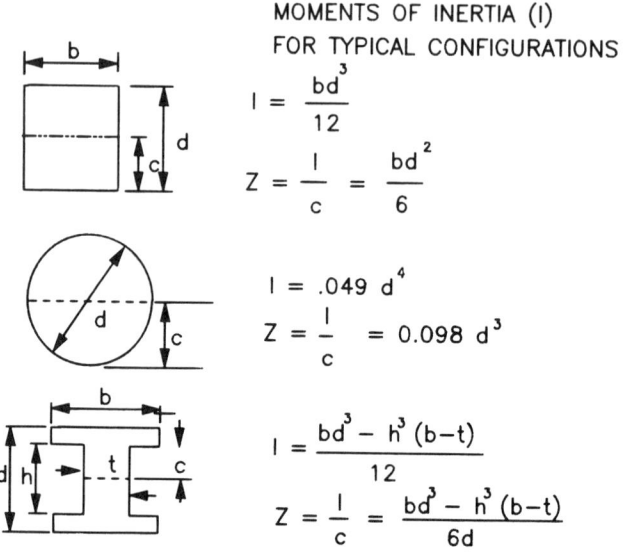

MOMENTS OF INERTIA (I)
FOR TYPICAL CONFIGURATIONS

$$I = \frac{bd^3}{12}$$

$$Z = \frac{I}{c} = \frac{bd^2}{6}$$

$$I = .049\, d^4$$

$$Z = \frac{I}{c} = 0.098\, d^3$$

$$I = \frac{bd^3 - h^3\,(b-t)}{12}$$

$$Z = \frac{I}{c} = \frac{bd^3 - h^3\,(b-t)}{6d}$$

MAXIMUM STRESS = S_{MAX}

DEFLECTION = Y_{MAX}

E = FLEXURAL MODULUS

I = MOMENT OF INERTIA

Z = SECTION MODULUS

Figure 15
Maximum stress and deflection equations for selected beams.

possible, while the designer often wants to restrict the draft due to a
functional or aesthetic requirement. Usually, a compromise will be required with a realistic value given for this parameter.

If the part is being textured and you plan to do this in the mold, draft
is an even more important factor. On a vertical wall you must have
at least 1 degree of draft for ever 0.001 for an inch of texture put into
the mold surface.

Ignoring this rule will result in a part that cannot be removed from the mold, or at best, a wearing away of the textured mold surface due to rubbing of the tool wall by the part as it is ejected from the mold.

The amount the part wall will increase due to draft can easily be calculated (see Table 1). The rule of thumb is that for every inch of vertical wall, the wall section should be increased by 0.017 in. for every degree of draft applied. It will not take many degrees of draft before the wall section will increase enough to cause a sink problem on the wall opposite the vertical rib.

MATERIAL SELECTION PROCESS

Selecting the right material is the next problem facing the design engineer. This may be influenced by the process that you have chosen, but since nearly all materials can be used in all processes, it is a good idea to narrow selection down to the specific material you plan to use. To determine the right material, you should answer all the questions as outlined in Chapter 3. By doing this, you can be reasonably sure that the material you pick will do the job. It is often a good idea to pick several materials that will meet the same goals. Often, the cost of the material will become an important factor and you may need to find a less expensive material that will do the job. Sometimes there will be little difference between two materials and you could select either one. When requirements such as heat resistance, chemical resistance,

Table 1 Dimension Differences Versus Draft Angles

Draw depth (in.)	½	1	2	3	4	5
1	0.009	0.017	0.035	0.052	0.070	0.088
2	0.017	0.035	0.070	0.105	0.140	0.175
3	0.026	0.053	0.105	0.157	0.210	0.263
4	0.036	0.070	0.140	0.210	0.280	0.350
5	0.044	0.088	0.175	0.262	0.350	0.437
6	0.052	0.105	0.209	0.314	0.420	0.525
7	0.061	0.123	0.244	0.367	0.490	0.612
8	0.070	0.140	0.279	0.419	0.059	0.700
9	0.078	0.158	0.314	0.472	0.629	0.787
10	0.087	0.175	0.349	0.524	0.699	0.875

or physical properties are taken into consideration, the material selection job becomes easier and only a few materials will really fit the specific need.

Material Information.   The material information required to do a good job in material selection is scattered in many places. Material suppliers and trade journals are generally the best place to obtain this information. Molders and moldmakers are also valuable sources of information. The important thing is not to take the first information you come upon as necessarily the only correct data on the subject. Sometimes there are two or more answers to a question and it is up to you to decide which is most correct for your application.

For example, General Electric offers a computerized data bank that goes beyond furnishing property sheets on their various materials. Through this service, one can find anything that he or she might want to know about a supplier's material and get some design help as well. Most material suppliers have some type of design assistance available in addition to providing information on their various materials. One should be careful not to take just one supplier's advice. There are often two materials available that will do the job and you have to decide which is right. Before the selection is made, it should be discussed with the molder.

COMPUTER-AIDED DESIGN

One technological breakthrough that is making a significant change in the way plastic parts are designed and tooled is computer-aided design (CAD). By having your part designed on a computer screen rather than a drafting board, you will have moved into the area of modern technology. This technique still has a few bugs, but many companies are adopting this method and it is just a matter of time before most parts will be developed on a computer monitor. Not only can one develop a completely engineered design of an object but can get any view, from any angle, as well as showing how the part will work when fitted into place with its mated counterparts. It takes a great deal of the guesswork out of designing a part and the system into which it fits.

With some software packages, after designing a part on a computer the designer can develop what is known as a data set. This is a series of points, lines, and surfaces defined in computer language and readable through another computer. The data set can then be used to give directions to an automated (NC) milling machine, which will machine the cavity and core, producing the mold from which the plastic part will be molded. By using this system, one can also go right into machining a tool-print model of the designed part. This technology involves a great deal of programming and computer skill, and in the early stages mistakes will probably be made. However, as personnel get more familiar with the method and equipment, more designers, molders, and toolmakers will adopt these new systems.

In the beginning of technological changes of this magnitude, one can only be as knowledgeable as possible about the new methods and be prepared for problems associated with them. The long-term benefits will be many; indeed, it is difficult to predict how far this technology will take us. However, there is still no substitute for experience and moldmaking skill. With all its bells and whistles, the computer is still simply a tool, and it needs the direction of a human being to make it functional.

CHOOSING THE RIGHT PROCESS

Once you have determined two or three materials that will work in your application, you are ready to choose the process to be used in producing the parts. First, the size of the part must be taken into consideration. If the part is too large, you may not be able to afford the tooling for standard injection molding. Structural foam may be a consideration if you can accept a swirl finish. With molding pressures reduced, you may be able to produce the item using aluminum tooling on a multinozzel structural foam machine or on a standard injection molding machine using a chemical blowing agent.

Wall Sections. If control of wall sections is not too important, you may be able to consider vacuum forming. With this process you can utilize less expensive tooling, and size is not too important. Very large parts are possible, and some exciting things are being done to control wall sections and to create rib structures and other features. The process is still rather labor intensive.

Container Shapes. When the application is in the shape of a container, two processes are worth considering: blow molding and rotational molding. Larger parts are usually rotational-molded, while smaller parts needing careful control of wall sections are generally blow-molded.

Cellular Products.   Products with cellular cores can be made by several different processes. Expanded polystyrene is often used for products of a cellular nature, and polyurethane can be produced in foam form using a process known as reaction injection molding (RIM). As mentioned earlier, structural foam also makes possible a product having cellular cores and integral skins.

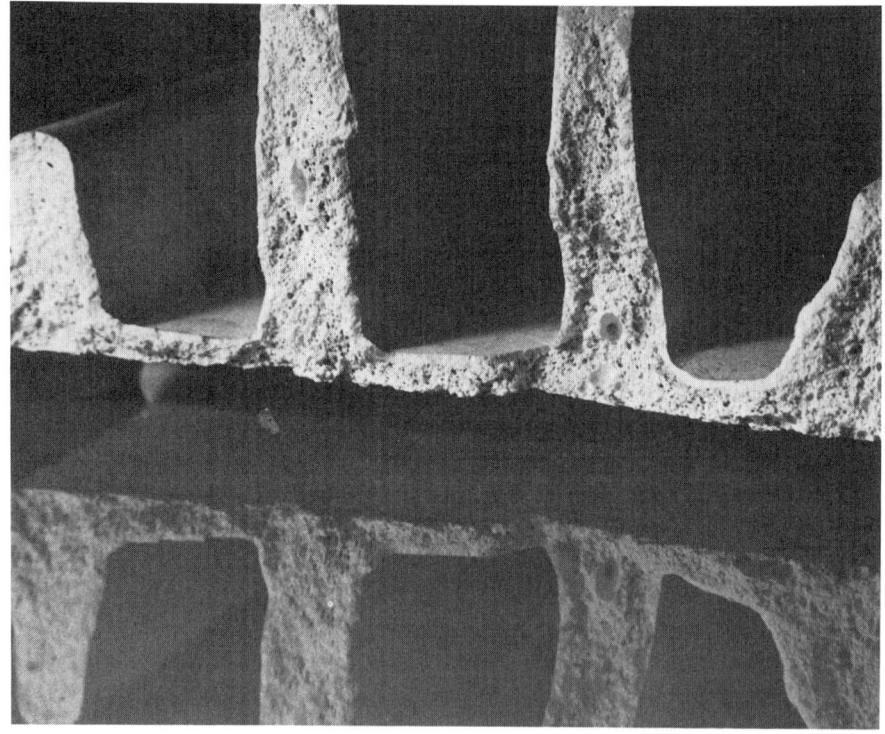

Figure 16
Structural foam showing integral skins and cellular core. (Photo courtesy of S.P.I.)

THE LOWEST BIDDER

Do not pick the lowest bid unless you are sure of the vendor and his capabilities. Molders usually quote in a similar way, and prices for parts should not be too far apart. You may find wide variations in the tooling quote, but if you have made sure that each molder is quoting on the same class of tooling, even these prices should not be that far apart. Make sure that the materials of construction are the same and that the vendor has not forgotten such things as required polishing or engraving that you have specified be included in the mold.

You should next put the cost information on a spread sheet so that you have all of the information available and can examine it all simultaneously. If you find that one quotation is exceptionally higher or lower than the majority of the bids, you may wish to go back to the vendor and ask them to review the numbers. If the vendor has made a calculating error, such a gesture will be well received.

CRITICAL DECISIONS

Of the decisions that you have to make on any project, the one that involves the most capital expenditure is the one involving tooling up for production of your product. The decisions that lead up to this are important, and some involve outlays of substantial cash. Each is important in the natural development of your product, but until you have reached the major decision as to whether to build tooling, you can take your losses and not be badly hurt either in the ego or financially.

When you have reached the tooling decision, there are often several choices to be made. You may find that there are several methods of construction open to you as well as several materials from which to choose. For example, you might produce the item in metal, concrete, or wood; or you might find plastics to be the best material to use. By this stage in the project you should be able to narrow this down to one or two selections. Often you will have to develop cost figures before you can make the right decision. The cost of tooling will play an important part in this. You may also be able to fabricate the product out of wood, metal, or plastic in the initial stages before you have the expense of tooling up. Before you make this decision you should consider the timing, cost, and volume of product you expect to need.

If you decide to build hard tooling to produce the product from plastic, you must be in a position to choose the material you wish to use. Shrinkage values are different for various materials, and you need to decide on a material of construction before the moldmaker can proceed. Once this decision is made, it is expensive to change the tool for a new shrinkage factor.

PROJECTED VOLUMES

The projected volume of the part is an important consideration and is often considered the most difficult to develop. You may want to start small and produce the product in one way, and then as the demand increases, switch to a more economical method. The projected volume for a particular application is often one of the most difficult numbers to come up with. In a small company where you make the decisions yourself, you can often gamble on the projected volume. However, in a larger company, usually no one wants to predict how many of anything the company is going to sell. Often, a great deal of time is spent in developing information of this type. The engineer is left waiting while market analysts try to determine the magic numbers. In the end it is usually a "gut"- feeling decision made by upper management.

WORKING WITH YOUR VENDOR

Once you have picked a vendor, work closely with him and keep him well informed of any changes in the project. Discuss with him your packaging plans, your long- and short-range volume projections, and most important, your timing expectations. Allow enough time to get the project completed, and do not cut the time short just because your boss says that he or she has to have the parts in an unrealistic time frame. Putting excess pressure on a vendor will only force him into mistakes that will take up more of your time. Sometimes, a bonus for quick delivery will get a week or two cut off the original time, but be careful of this. If the moldmaker is any good, he should have known what their workload was and how long it should have taken to build the tool in the first place. Ask for a weekly tooling construction schedule from the molder and visit the tool shop at least once during construction of the tool.

AFTER PLACING THE TOOL ORDER

Once the order for a tool has been placed, you will have time to take care of the miscellaneous items that you have been putting off up until now. These could include:

Packaging. Both bulk and individual packaging in which you would like the product to be placed should be evaluated.

Test marketing. By using models and preparing some preliminary literature, you can use this time to get a better feeling for the acceptance of the product.

Material. Make sure that the molder has the correct material on hand for the first test run.

Secondary jigs and fixtures. These should all be discussed with your molder and ordered if they are going to be needed after the first mold tryout.

Color. Color matching should be reviewed at this time, so that when the mold is ready, you will have material of the correct color on hand.

Inserts and miscellaneous screws and hardware. These should all be on order and samples available for initial part development.

FIRST MOLD TEST

When the time comes for the first mold test, be prepared for some problems. Buying a tool to produce a complex molded part is like buying a piece of machinery. You should expect problems with the first parts produced by a new tool. The startup of a new mold is always a trying situation and you can expect nearly anything to happen. The product may not match the prints, or any of a number of other things can go wrong. You should expect the quality of the part to go up as both the machine operator and the quality control personnel gain experience molding the product.

As production gets started, watch inventories carefully. Often, new buyers of plastic products order too many items in their initial orders. An injection molding machine can produce large quantities of product in a very short period of time. Early dimensions may be off or it may be determined that another feature is needed after the first parts are available. If the product does not sell or if something has to be changed, you may find yourself with unwanted inventory. This is a good way to tie up cash and can defeat a project before it gets started.

FIRST ARTICLE INSPECTION

First article inspection should be done as soon as you have a representative sample of the molded or extruded article from the molder. As the buyer, you should perform the inspection, although it is often a good idea to have the molder perform an inspection as well and compare inspection reports. Make sure that all functional dimensions are correct. Sit down with the molder and the toolmaker and discuss any corrections that are needed. Do not be afraid to ask for changes or corrections that you feel are the moldmaker's responsibility. Most molders do produce parts that match the prints. If the dimensions are close enough and the part is functional, be reasonable and let the dimensions stand. Change the prints to agree with the part. Make sure that the molder can duplicate the parts and that you will get consistent dimensions on production. If the moldmaker has to make corrections to the tools, insist on another first article molding run after the tool has been corrected. It is not unusual for a mold to go back to the toolmaker two or three times after the first trial.

FIRST PRODUCTION PARTS

Once you have gotten a correct first article, it is time to start thinking about a first production run. Before doing this, however, you should be prepared to pay for the last half of the tooling process. This is usually due upon receipt of an approved first article, and most molders will not run any production until this bill has been paid.

QUALITY CONTROL

You should set up a quality control program with your molder at this time. Sometimes it is better to wait until after the first production run is complete and a representative sample of the product is available for study. Set realistic dimensional tolerances and aesthetic criteria and have all prints brought up to date so that both you and the molder will know what is expected.

JUST-IN-TIME DELIVERY

Requiring a just-in-time delivery program is becoming more and more common. The thing to remember is that somebody has to pay for the added storage and inventory requirements that such a system demands. You will no doubt end up paying more for your parts if you don't buy them in realistic volumes.

Allow your molder a chance to produce adequate and dimensionally-correct parts. Don't be too hasty with the threat of moving your tool if a few problems arise. All molders go through a learning curve and your project will be no expection. Give you molder a chance to work the bugs out of the mold and get used to manufacturing your product just the way you want it. Remember, if you change molders you will have to go through it all over again with your new supplier.

TOOL OWNERSHIP

Another thing to be sure of when entering into a project with a molder is who owns the tooling from which the parts are made. This includes the mold base into which a tool insert fits and such other things as fixtures and special tooling necessary for the production of the parts. It is a common practice in the plastics industry for customers to do their own tooling. This is usually decided in the early part of the contract discussion. Be sure to ask: Who owns the mold and any fixtures? Customers want to be able to move their tools from the initial vendors if the situation so dictates.

Most tooling is designed to fit on any molding machine of the proper size and layout. Some molders utilize a mold base and insert system and only sell customers the inserts to fit into their bases. If customers do want to move their molds to other molders they have to purchase other bases to go with their inserts. This costs extra time and money, and usually a lot of ill will. Customers should be sure they understand the complete tooling situation when they buy molds and know exactly what the ramifications are if they should decide to change molders.

FISHING LURE EXAMPLE (CONTINUED)

The fishing lure project went fairly well, at least in the early stages. Once the mold was complete, other problems developed. The packaging was found to be a very important factor from the point-of-sale standpoint. A blister package was chosen and a backup card was printed that was designed to attract fishermen to the product. Problems then developed over several small design features that were not considered in the original design. The mold had to be altered to take care of these, and since these features had not been included in the initial drawings, a discussion ensued as to who should pay for the mold revisions.

It is important to ensure that everything done on a project be agreed to in writing. This is the way to eliminate problems from developing and to keep a project moving. Difficulties arising over verbal agreements and loose procedures will haunt a program and cause no end of problems. No one should feel that he or she is being too picky by insisting on clear written procedures and good engineering drawings before proceeding with a project. It is just good business.

CONTROL

Sometimes a project can be influenced by there being too much control. This is often the case when a large company is involved. Having too many specifications and policies can often hinder a project and cause it to run on longer than it should. In addition, it generally costs more money when this happens. Somewhere between overcontrol and undercontrol is a happy medium where both the manufacturer and the engineer can work together to accomplish what is needed.

3

Material Selection

In this chapter we cover the selection of the best polymer for a specific job. Also discussed are the physical tests that can be used to pick the correct plastic and a series of charts comparing these physical tests for a number of polymers. The tests include heat distortion, notched Izod impact, tensile strength, and flexural modulus. We also discuss application areas and which polymers are best suited for use in these areas.

MATERIAL SELECTION

Picking the best polymer to use in a plastic application is becoming increasingly difficult. The introduction of many new materials, and alloys of these materials and the more common polymers, has opened up a myriad of possibilities. One should also realize that information on plastic materials has been available for only a relatively short period of time. Unlike metals, the data on plastics are far less accurate, and wide variations can be found between reporting laboratories. Processing conditions can also alter the values reported by the basic material suppliers, and the addition of additives, including color pigment, can also cause properties to fluctuate.

Figure 17
A new 18.4-in. lawn mower with a plastic housing that offers abrasion, impact, and chemical resistance. (Photo courtesy of General Electric.)

THE RIGHT MATERIAL

In choosing the "right" material to do a job, you may find several products that will meet the requirements. They may vary in price or in processing characteristics, but in any case they present a choice. One should first list the conditions that the material must meet to fulfill the requirements of the application, then list all of the materials that provide the properties needed to meet this criteria.

The types of questions that must be answered are:

1. What is the highest temperature to which the application will be exposed during its life cycle?

2. What is the coldest temperature to which the application will be exposed?
3. How still must the material be to perform in the application?
4. Should the materials be transparent, translucent, or opaque?
5. What flammability resistance is required?
6. With what chemical environment will the application come into contact?
7. Will the application have to sustain any impact loading?

Once you have answered all seven questions, you should be able to choose one or several plastic polymers that will be able to perform in this application. You can then start considering such things as price, molding characteristics, and other factors that may help in selecting the best material.

If several of the materials that you choose have the same mold shrinkage, you may be able to delay your material decision until after the mold is built. However, if the materials have distinctly different mold shrinkages, you must pick one over the other before the mold is built, since the moldmaker will need to know to what shrinkage to build into the mold.

The tables provided later in this chapter will enable you to select the best material for a specific job. It should be noted that the data provided were taken from supplier's literature and may vary from one supplier to another. In all cases the material suppliers should be contacted regarding any material and asked whether it meets the requirements specified. Much design and material information is available from suppliers. The list of telephone numbers provided in Appendix III can be used to obtain other information about materials.

MATERIAL APPLICATION AREA

Each material has its own set of properties and attributes. Some materials are better in high-heat environments, and some materials are extremely abrasion resistant. The difficult part is to find a material that will come closest to fulfilling all of the requirements for your purposes. Following is a short summary of the outstanding features of some of the more popular plastic materials.

Polyethylene. This material is a member of the polyolefin family and is widely used in packaging and toy applications. It is probably the least expensive of the so-called commodity resins, but because of its excellent outdoor resistance and general toughness, it should not be overlooked for use on more critical engineering applications. The low-density form is used for packaging applications in such items as bottles and other containers. The higher-density material is found in such applications as toys, sporting goods, battery cases, and hundreds of other products. The material has excellent resistance to most solvents and chemicals and is impervious to moisture. It is especially good for outdoor applications and has good resistance to cold temperatures. The material comes in a wide range of molecular weights and hardness. It is easily processed on nearly every type of plastic processing equipment. It can be colored easily.

Polypropylene.   Another of the polyolefin family, this material is widely used where chemical and solvent resistance is important. It has good high-temperature resistance and is almost always chosen where a molded-in integral hinge is designed into a product. The material is often used for office equipment, such as trays, pencil holders, chair backs, and material handling containers of all kinds. The material is naturally translucent and can be colored easily.

Styrene.   Another commodity material, styrene is widely used by itself or in combination with other polymers. It is naturally clear, and a grade known as "crystal" styrene is used for optical applications where impact strength is not too important. The resin has a fairly high modulus and although short on impact resistance, is often used for such inexpensive items as swizzel sticks and drinking cups. When mixed with quantities of rubber, the material is called "high-impact" styrene and is often used in applications where high-temperature resistance is not needed. The material is used a great deal in toy applications, furniture parts, and in various eating utensils. It can be processed into structural foam or expanded polystyrene (EPS), where pentane gas is used to expand the styrene into a cellular product. This is used in foam cups, plates, and hundreds of other packaging applications.

ABS.   This popular material is really an alloy of three basic materials: acrylonitrile, butadiene, and styrene. The material has high strength, is easily colored, and has moderate temperature resistance. You will find it used in such well-known applications as telephone housings, furniture parts, and complicated housings of all types. If is also used for various electrical components where high-temperature resistance is not needed. It is available in a transparent version, but it is normally used in opaque form. Glass fibers are often added to stiffen the material.

Acrylic. This material is normally used where a transparent resin is required. Not high in impact strength, this material is often used in optical applications such as viewing windows, lenses, and lighting fixtures of all kinds. It can be colored easily.

Nylon. This well-known material comes in many grades and combinations. Translucent in its natural form, the material can be colored easily. Known for its extreme toughness, this resin finds its way into many industrial applications and electrical components. It does take on moisture in its molded form, which causes some dimensional problems, but this can usually be cured by soaking the molded part in water until it reaches an equilibrium state. For this reason, it is often used with concentrations of glass or carbon fiber. It has fair temperature resistance and good chemical resistance. A form of nylon is available for use on a RIM machine and so can be molded from a liquid polymer state.

Polycarbonate. Known for its high-temperature resistance (200 °F use temperature), this material finds its way into many tough industrial applications, such as housings, covers, and electrical components. It is naturally transparent, so it can be used in safety glasses, lighting fixtures, and glazing. It is more expensive than ABS and has limited solvent resistance. It can be alloyed with other materials, such as ABS and polyester, to form a completely different family of materials.

Modified Polyphenylene Oxide. Usually, an alloy of styrene and polyphenylene oxide, this material is used in electronic housings and similar applications. It has moderate temperature resistance and good physical characteristics.

Acetal. This material is used in many applications where toughness, chemical resistance, and good temperature resistance are required. It is naturally self-lubricating and because of this, is often used for gears and similar applications.

Polyester (Thermoplastic). This is one of the newer materials to become available. Most widely used in blown beverage bottles, its now being alloyed with other materials to form a new family of engineering materials with exceptional properties. It has extreme toughness and excellent heat resistance.

Thermoplastic Elastomers. This is a separate family of materials that include such generic resins as urethanes and butadiene–sytrene. These materials have excellent elongation and general overall toughness. They are often used in place of more conventional rubber. They are processed on standard plastic's machinery.

Polyamide imide, Polyphenylene sulfide, Polyether ether ketone, Polyether imide, Polyarylsulfone, Polysulfone, and Polyarylate. These are all polymers developed mainly for high-heat applications in recent years. Used in combination with glass or carbon fiber, they are finding great interest as a replacement for metal. Used in this form they are known as *composites* are processed in several ways. Applications include internal parts of the combustion engine, air craft components, and other demanding uses. Although still relatively expensive, these materials have been coming down in price as the industry finds more uses for them. They are often more difficult to process and usually require higher processing temperatures and higher pressures.

PHYSICAL PROPERTIES

The decision whether to use one or the other of the available polymers is normally based on an evaluation of their individual physical properties and processing features. Comparisons of many of plastic materials are provided in the tables that appear later in this chapter. These tables compare materials by generic name and indicates which are highest and lowest in several important properties including (1) heat deflection temperature, (2) impact strength, (3) flexural modulus, and (4) material cost. Also included is an explanation of how these physical properties are derived and why each is important in choosing a particular plastic.

DESCRIPTION OF IMPORTANT PHYSICAL TESTS

To understand the importance of physical properties used with plastics, an explanation of important tests and terms is given below.

Tensile Strength. The word *tensile* means to pull apart. Tensile strength is the resistance of a material to being pulled apart. This is usually expressed in pounds per square inch. One square inch of cottom would require little force to pull it apart. However, to pull plastic apart may require from 1000 to 50,000 psi. Steel and other metal can run as high as hundreds of thousands of pounds per square inch.

Elongation. This property is always associated with tensile strength and is the increase in original length at point of fracture. This is expressed as a percentage. If we take a piece of rubber 4 in. long and can stretch it to a length of 12 in. before it breaks, we say that its elongation is 200%.

Tensile strength and elongation are also important where toughness is required. A material that has high tensile strength and relatively high elongation is a tougher material than one having high tensile strength and low elongation.

Modulus. This term may be applied to tensile, compressive, flexual, or torsional action. It defines the number of pounds per square inch required to cause deformation, elongation, flexure, and so on, in a material. In other words, it represents stiffness. Consider a nylon bearing with a ½-in. wall, which is to support a load of 2000 psi. One of the considerations is how much additional clearance will be developed due to the elasticity of the bearing. In this case, the modulus of the material is found by dividing the load by the resulting deformation, which is expressed as a percentage. The modulus of nylon in compression is approximately 400,000 psi. To find the percentage of deformation, divide the 2000-psi load by 400,000; the answer is 0.5%. Multiply this by the ½-in. wall thickness and we find that a deformation of 0.0025 in. will occur.

Flexural Strength. This property, which is expressed in pounds per square inch, is an indication of stiffness. A material will have a higher flexural strength value if it supports its own weight. This property is important in applications where the plastic is being bent or continually flexed.

Hardness. There is no exact term that defines hardness. It is expressed in some cases by the indent of a small ball, and in other cases by the penetration of a sharp spring-loaded point of metal. Generally, a harder surface provides better wear and abrasion resistance. However, this may not be the case in materials such as elastomers, where the surface backs away from the abrading force and the abrasion resistance is high.

Compressive Strength.   This property measures the maximum load in pounds that a 1-in. section of material will support without fracturing. It is a good property to use when comparing two materials, but after selecting the one with the higher value, other considerations should be used. In other words, this test should not be the only test used for a given material.

Coefficient of Linear Thermal Expansion.   This property deals with the amount of material will grow when it is heated, normally expressed in inches per degree. The thermal expansion for plastics is four to eight times higher than for other engineering materials and is therefore an important consideration.

Heat Distortion.   This property is used primarily to compare different materials. It is the temperature at which a material bends a predetermined amount under a given load.

Moment of Inertia.   Geometric sections are used to find the moment of inertia. Each is specific to the cross section of the material shape under consideration. The symbol for moment of inertia is I. *Hooke's law*, the main equation for strength of materials, is

$$\text{modulus} = \frac{\text{stress}}{\text{strain}}$$

$$\text{or } E = \frac{S}{\epsilon}$$

and is dependent on the material being used.

The *bending stress formula* for analyzing stress in a flexurally loaded beam is defined by the equation

$$S = \frac{M}{I/c} = \frac{Mc}{I}$$

where

S = bending stress
M = bending moment, determined by the length, supports, and loading of the beam
I = moment of inertia
c = distance from neutral axis to extreme outer fiber

The dimensions involved and information for a specific material will be known for each configuration.

All of these properties should be used with caution; they are primarily a means of comparing one material with another. Although there are several design formulas that can be used to calculate whether a given load can be accommodated, nothing is better than an actual test on a prototype.

Table 1 lists physical properties, trade names, and manufacturer's of various materials. The list is not meant to be complete but gives only a representative sampling of trade names and manufacturers.

Table 2 provides a comparison of heat distortion temperatures for various materials.

Table 3 compares tensile strength values for various polymers.

Table 4 compares flex modulus values for various materials.

Table 5 compares Izod impact values for various materials.

ELECTRICAL PROPERTIES

Electrical properties are important for all applications where electrical current is involved. Whether dealing with the cover for a hand-held drill or an extruded insulation on an electrical wire, the electrical properties must be taken into consideration.

The properties to be evaluated when looking at plastic as an insulator are the following:

Dielectric Strength. This is a measure of the electrical breakdown resistance of a polymer under an applied voltage stress. It is tested using ASTM D149. Temperature, preconditioning, and the thickness of the material can all affect dielectric strength.

Table 1 Physical Properties for a Variety of Polymers

Generic name	Trade name	Manufacturer	Heat distortion temp. at 264 psi (°F)	Notched Izod impact, ⅛ in. at 72°F	Tensile strength at yield (psi)	Flex modulus (psi)
Acrylonitrile–butadiene–styrene (ABS)	Lustran	Monsanto	185	6.2	6,100	380,000
	Cycolac T	Borg-Warner	210	6.5	6,000	340,000
ABS–polycarbonate blend	Bayblend DP21011	Mobay	223	1.9	4,100	819,000
Acetal (homopolymer)	Celcon	Celanese	225	1.3	8,800	375,000
Acetal (copolymer)	Delrin 100 nc 10	Du Pont	277	2.5	10,000	415,000
Acrylic	CP51	Continental Polymers	190	0.5	10,000	13,000
	Plexiglas M-17	Rohm & Haas	190	0.6	7,000	350,000
Butadiene–styrene	Stereon 880	Firestone	153	0.3	4,850	240,000
Cellulosics (butyrate)	Tenite S2-H4	Eastman	196	1.4	6,100	280,000
Fluorocarbon	Teflon FEP 100	Du Pont	250	2.9	3,00	95,000
Liquid polymer	Xydar SRT 500	Dartco Mfg.	639	3.9	18,200	1,900,000
Methyl methacrylate–styrene copolymer	CTE	Richardson	210	0.4	7,100	390,000

Nylon–ABS blend	Triax 1125	Monsanto	198	16.0	6,800	310,000
Nylon 6	Zytel 211	Du Pont	180	1.5	12,000	30,000
Nylon 6, 12	Zytel 158 L	Du Pont	194	1.0	8,800	295,000
Nylon 6, 6	Zytel 101	Du Pont	167	1.0	7,500	160,000
Nylon 11	Rilan 11	Meachem	131	4.4	8,250	142,000
Nylon 6, 10	Ultramid 800 BC	EBF Corp.	194	1.0	8,800	295,000
Polyallomer	Tenite 5021	Eastman	123	0.5	4,980	NA
Polyamide imide	Torlon	Union Carbide	525	2.5	27,000	664,000
Polyarylate	Arylon 401 NC—10	Du Pont	311	5.4	12,000	300,000
	Durel 400	Celanese	340	5.5	10,000	330,000
Polyarylsulfone	Radel A400	Union Carbide	400	1.6	12,000	385,000
Polycarbonate	Lexan 101	General Electric	270	17.0	9,000	340,000
	Makrolon PC 3200	Mobay	270	18.0	9,300	333,000
Polycarbonate–PET polyester	Makroblend dp4135135	Mobay	252	2.4	13,000	850,000
Polyester PBT (thermoplastic)	Petlon 4530	Mobay	438	1.7	19,700	1,490,000
Polyether ether ketone	Peek	ICI	320	1.6	14,500	550,000
Polyether imide	Ultem 1000	General Electric	392	1.0	15,200	480,000

(continued)

Table 1 (continued)

Generic name	Trade name	Manufacturer	Heat distortion temp. at 264 psi (°F)	Notched Izod impact, ⅛ in. at 72 °F	Tensile strength at yield (psi)	Flex modulus (psi)
Polyether sulfone	Victrex 300P	ICI	203	1.6	12,200	373,000
Polyethylene (low density)	Norchem 1028	Enron Chemical	111	1.0	2,200	45,000
Polyethylene (high density)	240 B-2	Amoco	172	1.0	3,800	155,000
Polyether imide (glass reinforced)	Ultem 2100	General Electric	410	1.1	16,600	650,000
Polymethylpentene	TPX RT-18	Mitsui	194	1.5	3,357	185,715
Polyphenylene oxide	Noryl N-190	General Electric	190	7.0	7,000	325,000
Polyphenylene sulfide	Ryton R-4	Phillips	500	1.3	17,500	1,700,000
Polypropylene (homopolyer)	5610.0	Shell	180	0.5	5,000	210,000
Polypropylene (copolymer)	Profax 7323 A	Hercules	147	2.2	3,800	175,000

Polypropylene (glass reinforced)	P-30fg-0100	Thermofil	275	1.1	7,000	700,000
Polypropylene (calcium carbonate reinforced)	Polyfil C-40	Polifil Inc.	235	1.0	3,400	320,000
Polysulfone	Udel P-1700	Union Carbide	345	1.3	10,200	390,000
Polyurethane elastomer	Isopast 101	Upjohn	194	22.0	7,600	235,000
Polyvinylidene fluoride	Kynar PVDF	Pennwalt	194	8.0	7,400	260,000
Propionate	Tenite 306 H-5	Eastman	188	0.9	6,300	305,000
Styrene	Styron XL 8028	Dow	160	3.1	2,700	300,000
Styrene–acrylon–nitrile (SAN)	Lustran	Monsanto	220	0.6	12,000	560,000
Styrene–maleic-anhydride terpolymer	Cadon 140	Cadon	262	2.0	5,200	340,000
Styrene–polycarbonate blend	Arloy 1100	Arco	257	10.0	7,000	335,000
Vinyl	Geon 87300	B. F. Goodrich	63 .5	17.1	6,400	361,000

Table 2   Heat Distortion Temperatures for Various Polymers[a]

Generic name	Heat distortion temp. at 264 psi (°F)
Vinyl	63.5
Polyethylene (low density)	111
Polyallomer	123
Nylon 11	131
Polypropylene (copolymer)	147
Butadiene–styrene	153
Styrene	160
Nylon 6,6	167
Polyethylene (high density)	172
Polypropylene (homopolymer)	180
Nylon 6	180
Acrylonitrile–butadiene–styrene	185
Proprionate	188
Polyphenylene oxide	190
Acrylic	190
Nylon 6,10	194
Polyurethane elastomer	194
Nylon 6,12	194
Polymethylpentene	194
Polyvinylidene fluoride	194
Cellulosics (butyrate)	196
Nylon–ABS blend	198
Polyether sulfone	203
Acrylonitril–butadine–styrene	210
Methyl methacrylate–styrene copolymer	210
Styrene–acrylon–nitrile (SAN)	220
ABS–polycarbonate blend	223
Acetal (homopolymer)	225
Polypropylene (calcium carbonate reinforced)	235
Fluorocarbon	250
Polycarbonate–PET polyester	252
Styrene–polycarbonate blend	257
Styrene–maleic–anhydride terpolymer	262
Polycarbonate	270

continued

Table 2 *Continued*

Generic name	Heat distortion temp. at 264 psi (°F)
Polypropylene (glass reinforced)	275
Acetal (copolymer)	277
Polyarylate	311
Polyether either ketone	320
Polyarylate	340
Polysulfone	345
Polyether imide	392
Polyarylsulfone	400
Polyether imide (glass reinforced)	410
Polyester PBT (thermoplastic)	438
Polyphenylene sulfide	500
Polyamide imide	525

[a]Materials tested using ASTM D648.

Table 3 Tensile Strength for Various Polymer[a]

Generic, name	Tensile Strength at yield (psi)
Methyl methacrylate-styrene copolymer	2,000
Propionate	2,700
Acrylic	3,000
Polypropylene (homopolymer)	3,357
Vinyl	3,400
Polyvinylidene fluoride	3,800
Polymethylpentene	3,800
Butadiene-styrene	4,100
Polyarylsulfone	4,850
Nylon 11	4,980
Nylon 6,10	5,000
Liquid polymer	5,200
Acrylonitrile-butadiene-styrene (ABS)	6,000
Nylon 6,6	6,100
Acrylic	6,100

continued

Table 3 *Continued*

Generic, name	Tensile Strength at yield (psi)
Polyamide imide	6,300
Polysulfone	6,400
Acetal (homopolymer)	6,800
Styrene	7,000
Cellulosics (butyrate)	7,000
Acrylonitrile–butadiene–styrene (ABS)	7,000
Polyether imide (glass reinforced)	7,000
Polyether imide	7,400
Styrene–polycarbonate blend	7,500
Polycarbonate	7,600
Fluorocarbon	8,100
Polyarylate	8,250
Polyurethane elastomer	8,800
Polyarylate	8,800
Nylon 6,12	8,800
ABS–polycarbonate blend	9,000
Nylon–ABS blend	9,300
Styrene–maleic–anhydride terpolymer	10,000
Polypropylene (glass reinforced)	10,000
Polester PBT (thermoplastic)	10,000
Polyethylene (low density)	10,200
Polyethyulene (high density)	12,000
Polyether sulfone	12,000
Polyether ether ketone	12,000
Polyphenylene oxide	12,000
Acetal (copolymer)	12,200
Polypropylene (calcium carbonate reinforced)	13,000
Polyallomer	14,500
Polyphenylene sulfide	15,200
Polypropylene (copolymer)	16,600
Nylon 6	17,500
Polycarbonate–PET polyester	18,200
Styrene–acrylon–nitrile (SAN)	19,700

[a]Materials tested using ASTM D638.

Table 4 Flex Modulus for Various Polymers[a]

Generic name	Flex modulus (psi)
Acrylic	13,000
Nylon 6	30,000
Polyethylene (low density)	45,000
Fluorocarbon	95,000
Nylon 11	142,000
Polyethylene (high density)	155,000
Nylon 6,6	160,000
Polypropylene (copolymer)	175,000
Polymethylpentene	185,715
Polypropylene (homopolymer)	210,000
Polyurethane elastomer	235,000
Butadiene–Sytrene	240,000
Polyvinylidene fluoride	260,000
Cellulosics (butyrate)	280,000
Nylon 6,12	295,000
Nylon 6,10	295,000
Styrene	300,000
Polyarylate	300,000
Propionate	305,000
Nylon–ABS blend	310,000
Polypropylene (calcium carbonate reinforced)	320,000
Polyphenylene oxide	325,000
Polyarylate	330,000
Polycarbonate	333,000
Styrene-polycarbonate blend	335,000
Polycarbonate	340,000
Styrene–maleic–anhydride terpolymer	340,000
Acrylonitrile-butadiene-styrene (ABS)	340,000
Acrylic	350,000
Vinyl	361,000
Polyether sulfone	373,000
Acetal (homopolymer)	375,000
Acrylonitrile–butadiene–styrene (ABS)	380,000
Polyarylsulfone	385,000
Methyl methacrylate-styrene copolymer	390,000

continued

Table 4 *Continued*

Generic name	Flex modulus (psi)
Polysulfone	390,000
Acetal (copolymer)	415,000
Polyetherimide	480,000
Polyether ether ketone	550,000
Styrene–acrylon–nitrile (SAN)	560,000
Polyetherimide (glass reinforced)	650,000
Polyamide imide	664,000
Polypropylene (glass reinforced)	700,000
ABS–polycarbonate blend	819,000
Polycarbonate–PET polyester	850,000
Polyester PBT (thermoplastic)	1,490,000
Polyphenylene sulfide	1,700,000
Liquid polymer	1,900,000

[a]Materials tested using ASTM D790.

Table 5 Notched Izod Impact Values for Various Polymers[a]

Generic name	Notched Izod 1/8 in. at 72 °F
Nylon 6,6	0.3
Acrylonitrile–butadiene–styrene (ABS)	0.4
Butadiene–styrene	0.5
Nylon 6,12	0.5
Polypropylene (glass reinforced)	0.5
Polyester PBT (thermoplastic)	0.6
Styrene	0.6
Polyarylsulfone	0.9
Acrylic	1.0
Polycarbonate–PET polyester	1.0
Fluorocarbon	1.0
Acetal (homopolymer)	1.0
Polyether ether ketone	1.0
Polyurethane elastomer	1.0
Styrene–polycarbonate blend	1.1

continued

Table 5 *Continued*

Generic name	Notched Izod 1/8 in. at 72 °F
Polyarylate	1.1
Nylon 11	1.3
Polycarbonate	1.3
Polyethylene (high density)	1.4
Styrene–maleic–anhydride terpolymer	1.5
Polyphenylene oxide	1.5
ABS–polycarbonate blend	1.6
Nylon–ABS blend	1.6
Polypropylene (calcium carbonate reinforced)	1.6
Styrene–acrylon–nitrile (SAN)	1.7
Polyallomer	1.9
Polyphenylene sulfide	2.0
Acetal (copolymer)	2.2
Methyl methacrylate–styrene copolymer	2.4
Polymethylpentene	2.5
Polypropylene (copolymer)	2.5
Polypropylene (homopolymer)	2.9
Polyether imide (glass reinforced)	3.1
Nylon 6	3.9
Nylon 6,10	4.4
Polyvinlylidene fluoride	5.4
Cellulosics (butyrate)	5.5
Vinyl	6.2
Polyethylene (low density)	6.5
Polycarbonate	7.0
Polyether imide	8.0
Polyamide imide	10.0
Propionate	16.0
Polyether sulfone	17.0
Liquid polymer	17.1
Acrylonitrile–butadiene–styrene (ABS)	18.0

[a]Materials tested using ASTM D25.

Volume Resistivity. This is a measure of the resistivity of electrical dc current through the thickness of a specimen, expressed in Ω-cm. Materials with resistivity below 10 Ω-cm are partial conductors. Materials with resistivities above 10 Ω-cm are considered insulators. This property is tested using ASTM D257.

Dielectric Constant. This is a measure of relative inefficiency, a dimensionless factor derived by dividing the parallel capacitance of the material by that of an equivalent volume of vacuum. This is tested using ASTM D150.

Arc Resistance. This is the time in seconds that an arc may play across the surface of a material without rendering it conductive. It is measured using ASTM D495 and is dependent on temperature, frequency, and conditioning.

Other tests are very important when evaluating the electrical properties of plastics. These include comparative tracking index (ASTM 3638-77), hot wire ignition (UL 746A, section 42), high current arc (UL 746A, section 43), and Underwriters' Laboratories flammability test (UL 94). Test information for most plastics is available from materials suppliers.

The electrical properties of various polymers are compared and listed in Table 6.

MOLD SHRINKAGE

The shrinkage of each material has to be taken into consideration before the moldmaker can hope to build a mold that will mold a part successfully. Ordinarily, it would just be a matter of taking the information supplied by the material manufacturer and using that. (Standard tolerances values for various polymers are provided in Appendix IV.) However, as technology has gotten more sophisticated, differences between shrinkage values have been noted for the same material flowing in different directions. That is, the shrinkage may be different for material flowing lengthwise with the gate than for material flowing crosswise to the gate. Careful consideration of this factor is important if a plastic part is to be produced within a close tolerance.

Table 6 Electrical Properties of Various Polymers

Polymer	ASTM D149 Dielectric strength (V/mil)	ASTM D257 Volume resistivity (MΩ-cm)	ASTM D150 Dielectric constant at 60 Hz	ASTM D495 Arc resistance (sec)
Acrylic	500	—	3.3	No track
ABS	377	9.3×10^7	—	86
Acetal(copolymer)	500	10^{15}	3.7	240
Acetal(homopolymer)	380	10^{15}	3.7	129
Nylon 6,6	530	10^{15}	4.0	—
Nylon 6	400	10^{15}	9.8	—
Nylon 12	449	2.9×10^{16}	3.5	120
Polyarylsulfone	383	3.03×10^{16}	3.51	81
Polycarborate	400	10^{16}	3.0	90
Polyether imide	580	10^{17}	3.1	140
PEEK	190	4.9×10^{16}	—	—
Polyester	470	4×10^{16}	3.3	63
Polyether elastomer	440	1.2×10^{11}	4.5	—
Polyphenylene oxide	400	10^{16}	2.65	75
Polyphenylene sulfide	450	5×10^{15}	4.3	116
Polypropylene	600	10^{16}	2.25	—
Polyvinyl chloride	787	7.5×10^8	3.9	7

Table 7 Some Characteristics of Common Plastics

	Specific gravity	Flame color (copper wire)		Color	Smoke density	Odor	Solvents[b]	Comments
		As is	Melts/Soft					
Polypropylene	0.85–0.9	Blue-yellow	Yes (trans)	White	Very little	Heavy	Toluene[b] (slowly slight)	Drips, swells
LDPE	0.91–0.93	Blue-yellow	Yes (trans)	White	Very little	Candle wax	Dipropylene glycol[b]	Drips, swells
HDPE	0.93–0.96	Blue-yellow	Yes (trans)	White	Very little	Candle wax	Toluene[b]	Drips, swells
Epoxy	1–1.25	Orange-yellow (green)	No	Black		Phenolic		Some soot
Chlorinated PE	1–2.4	Green	Yes				Tolune[b]	
Polystyrene	1.05–1.08	Orange-yellow	Yes	Black	Dense	Sweet marigolds	Diethyl benzene[b]	Soot, no drip
Polyvinyl butyral	1.07–1.08	Blue mantle, yellow,	Yes (trans)			Rancid butter		Drips, swells
Nylon	1.09–1.14	Blue mantle, yellow	Yes			Burned hair		Swells, froths
Ethyl cellulose	1.1–1.16	Blue-white	Yes			Sweet	Sec-Amyl alcohol	Drips

Material							
Polyester	1.12–1.46	Yellow	No	Black	Dense	Sweet (resinous)		Softens
Vinyl chloride	1.15–1.65	(Green) yellow-orange	Yes softening	White to green	Little	Acrid, chlorine	Toluene[b]	No drip
Arcylic	1.18–1.19	Blue mantle yellow-orange	Yes (trans)	Some black		Floral, burned fat	Toluene[b]	Clear bead
Vinyl acetate	1.19	Dark yellow	Yes	Black		Acetic	sec-Hexyl Alcohol, cyclohexanol, acetonitrile	Some swell
Polycarbonate	1.20	Orange-yellow	No	Black		Phenolic, sweet	Toluene[b]	Chars
Cellulose acetate	1.27–1.34	Dark yellow, mauve blue	Yes	Black		Acetic, vinegar	Furfuryl alcohol & acetonitrile	Burns, charred bead
Casein	1.35	Yellow	No	Gray		Burned milk		Swells, chars
Cellulose nitrate	1.35–1.40	Intense white	Yes			No odor	Dipropylene glycol & acetonitrile	

(continued)

Table 7 *continued*

	Specific gravity	Flame color (copper wire)		Color	Smoke density	Odor	Solvents[b]	Comments
		As is	Melts/Soft					
Acetal	1.41–1.42	Blue mantle, yellow	Yes			Formaldehyde		Drips
Urea–formaldehyde	1.47–1.52		No			Urinous		
Melamin–formaldehyde	1.50–220		No			Fish		
Phenol–formaldehyde	1.55–1.90		No			Phenolic		
Saran	1.58–1.75		Yes					
Vinylidene chloride	1.62–1.72	(Green) yellow	Yes			Sweet		Heavy black
Chlorinated rubber	1.64		Softens	Black	Dense	Rubber		
Alkyd	1.80–2.24		No					
Tetrafluoroethylene	2.1–2.3		No			Burned hair		Chars
Neoprene		(Green) orange	Softens	Black		Rubber		

Source:
[a]Test for halogen (chlorine).
[b]Hot.

Materials that incorporate glass or carbon fiber demonstrate greatly reduced shrinkage, but fiber orientation can affect the way the part shrinks in the mold. The only way to make sure that this problem will not affect a molded part is to utilize a computer program designed to calculate and compensate for the different values. It is extremely important that you pick a molder and moldmaker who have had experience with this problem and work closely with them.

Miscellaneous characteristics of a variety of plastics are presented in Table 7.

4

Modifying Plastic

WHY MODIFY PLASTIC RESIN?

Often there is a demand to change the plastic resin you plan to use in an application. Whether you do this at the molder by adding concentrates, glass or carbon fiber, or possibly something to improve lubricity, or purchase such modified polymers from a material supplier, the result is usually the same. Often, the modified resin that you need is not available through a material supplier. This is when the experience of the molder plays an important part.

Be sure that you check with your material supplier about the compatibility of various modifiers before using them. Some resins can be mixed, whereas others will show signs of delamination when they are used with incompatible modifiers.

Good sources of modified resins are companies that specialize in reprocessing polymers. Such companies have a wide selection of specialty plastics for sale, and you can be sure that the various combined resins are both mixed well and are compatible.

COLORING PLASTIC PARTS

Adding color is an important ingredient in any project. It can either make the project or keep it from selling. So it is obviously important.

There are many ways one can get color into parts, no matter what process you choose to produce a part. With the advent of the computer to match colors, it is possible to get an accurate color match for nearly every base material and plastic process.

Colored plastic resin can be obtained by using a precolored material, by adding color concentrate or dry color or by dyeing or painting. The precolored material is colored by the material manufacturer and is probably the most accurate and uniform of the coloring methods. However, quantities must be in the 500-lb range before a material supplier will be willing to bother with it.

Color Concentrates.   Color concentrates is another way to achieve colored parts. This technique uses a heavy concentration of color pigment put into a small amount of base resin. The concentrate is then mixed with natural resin by the processor.

The *let-down ratio*, the ratio of concentrate to natural resin, varies with the color and the base resin used. This is a good way to do small volumes of material since several concentrate companies will sell you as little as 5 lb of concentrate. Granted, the pound price is very high at these small volumes, but it is the most economical way to get good consistent color.

Dry Color.   Dry color and other forms of pure color pigment are added directly to natural resin by the processor. Because the amounts put in are not easily controlled, this method of coloring is very inaccurate. It is also extremely messy, and most processors try to avoid using it if they can. It is, however, a very economical way if great accuracy is not required.

Dyeing.   Another method used to color parts is to use a water-soluble dye and a base resin that absorbs some quantity of water. The part is molded in natural resin and then emersed in the dye, giving it a colored effect.

Painting. A part can also be painted using a durable coating such as Sherwin-Williams' Polane urethane. This is often the most expensive way to color a plastic part. More details on this subject are given later.

The one thing to be certain of no matter how you color your plastic part is to be in agreement on just what color you require. Colors can be difficult to see and are often seen differently by different people. Be sure you have seen and agreed upon a color chip made of the same material from which your part is made. Also, watch the thickness of the color chip used as a sample. In different materials the thickness of the part will strongly affect the apparent color.

COLOR SELECTION

Color selection for a plastic product may be important for other than aesthetic reasons. It is well known that the surface temperature of an object exposed to sunlight is dependent on its color. If the product is for exterior use, the selection of color may be critical from this standpoint. This is especially true in the case of products such as golf cart bodies or boats.

The everyday use of these types of products will involve considerable contact with a person's body. For these types of products it is essential that the surface temperature build-up be held to a level that feels comfortable to the touch. Table 1, which shows typical surface temperatures measured on various colored ABS panels, is an aid to the designer in making this choice for products made of Cycolac ABS. These measurements were made on panels exposed to an actual outdoor environment at an ambient air temperature of 100 °F. Airflow around the panels was unrestricted.

Estimation of the influence of surface temperature buildup should include consideration of the total product assembly. In applications where airflow around the part is restricted, surface temperatures will be somewhat greater. Also, to obtain added stiffness, some applications require that the Cycolac ABS be backed by urethane or polystyrene foam. In these cases, the excellent insulating characteristics of the foam materials result in even greater surface temperatures, as shown in Table 1.

Table 1 Surface Temperatures of Various Panel Colors[a]

	Surface temperature (°F)	
Panel color	Unbacked	Foam backed
White	120	127
Light blue	127	137
Medium blue	137	157
Dark blue	144	174
Medium red	128	142
Dark red	137	169
Black	148	173

Source:
[a]Ambient air temperature = 100 °F.

Warpage.   Another area in which surface temperature can play an important role is part warpage. All materials expand when subjected to an increase in temperature. The property that describes this characteristic is called the *material coefficient of linear, thermal expansion* and is expressed as the change in material length per unit per degree change in temperature.

Consider the example of two parts, one black and one white, which are identical with the exception of color. Based on the surface temperatures listed in Table 1, exposure of the parts to a sunny outdoor environment will result in greater expansion of the black part due to a greater surface temperature buildup. If expansion of the part is restricted in any manner, such as being attached to a framework of a material having a lower thermal coefficient, stresses will develop. Warpage, or distortion, is one of the ways in which the part will tend to relieve these stresses.

In summary, color, which directly influences surface temperature, has an influence on part expansion and potential part warpage. The data presented here shows that these potential problem areas may be minimized through the use of pastel colors.

Ultraviolet-Light Degradation.*   In addition to imparting color, pigments offer significant protection against the damaging effects of

---

*This information was supplied by PDI of Edison, NJ.

daylight. To plastics, the most destructive wavelengths of light are in the ultraviolet (UV) region, over the range 290 to 380 nm. UV light waves degrade plastic in two ways: by direct rupture of chemical bonds or by energy transfer, sensitized by UV-excited impurities.

The plastics and paint industries have long recognized that these types of UV degradation can be decelerated by the incorporation of pigments. Many pigments absorb UV radiation and act as screening agents. Others reflect and scatter light waves. In both cases, the UV protection offered by pigments is superior to that of light stabilizers which are consumed by UV light, leaving the compound unprotected.

Since light wavelengths between 290 and 380 nm are most harmful to the majority of polymers, the pigments selected must exhibit strong UV screening capabilities within this wavelength range. Not all color pigments exhibit the same ultraviolet screening capabilities. However, those with excellent screening qualities produce a wide range of finished blended colors.

Although pigments have no affinity for the plastics in which they are used, they do contribute to preserving a part's physical integrity. Cracking, checking, loss of adhesion, and loss of tensile strength are all substantially reduced when suitable pigments are properly molded into the plastic component.

MATERIAL SELECTION

Picking the correct polymer to be used in your plastics application is becoming an increasingly difficult task. The introduction of many new materials and the alloys of these new materials with the more common polymers has opened up a myriad of possibilities.

In choosing the "right" material to do your job, you may find several products that will do the job. They may vary in price or in molding characteristics, but in any case, present a question of choice. One should first list all the conditions that the material must meet. Then match these up with the materials that appear to be well suited for the application. The questions answered must include the following:

1. What is the highest temperature that the application must sustain?
2. What is the coldest temperature that the application must sustain?
3. How stiff must the material be to perform in the application?
4. Should the material be transparent, translucent, or opaque?

5. What flammability resistance is necessary?
6. With which chemical environment will the material come into contact?

Once you have answered all six of these questions, you should be able to pick a list of several plastic polymers that will be able to do the job. Then you can start considering such things as price and molding characteristics. If a couple of the materials you end up with have the same mold shrinkage, you may be able to delay your decision until after the mold is built. However, if two materials picked have distinctly different mold shrinkages, you must pick one of them before the mold is built, since the moldmaker will have to know what shrinkage to build into the mold. You can use the tables in Chapter 3 to help you in selecting the best material to do a specific job.

Much design and material selection information is available from various material suppliers. A list of selected telephone numbers is given in Appendix III.

FINISHING

Sometimes it becomes necessary to apply a coat of paint to a plastic part. This can be because the surface is rough, as in a structural foam application, or to cover knit and weld lines in an injection-molded part. In the case of structural foam, the part usually gets coated first with a primer or barrier coat of water-based acrylic. This protects the surface so that the solvents in the urethane outer coating will not attack the base resin. The next coat to be applied is the color coat of urethane. This is followed by a stipple coating to add texture to the part. The overall thickness is about 6 mils. This provides an excellent protective coating over the surface of the part, giving it both a protective barrier against chemical attack and a tough, hard-to-scratch surface to fight abrasion. These properties, given to a plastic part of any kind, are often worth the extra money spent on finishing. On top of that, you have solid color control, a feature sometimes hard to come by in a molded part.

In the case of coating on a solid injection-molded application, one can often cover molding imperfections, such as knit and weld lines as well as minor sink marks that often plague molded parts. Again you can get perfect color control, and often, it takes only one coat of paint. More often than not, urethane coatings are chosen for this type of

Figure 18
Many plastic parts require a coating of urethane paint to provide color, scratch resistance, and a protective surface. (Photo courtesy of Xytec.)

application. A coating such as Sherwin-Williams' Polane systems are often specified because of their ease of application and generally good physical properties.

The amount of secondary necessary to prepare a plastic part for finishing is often very expensive and time consuming. Often, the part must be thoroughly sanded and sometimes a plastic putty is needed to fill large holes or depressions. This must be sanded down so that the surface appears flat. Also, inserts and other areas that one does not want coated with paint must be masked off. This can be expensive if there are 20 or so inserts in the final part. When stippling a finished part, it is often necessary to mask off areas that you do not want covered with the stipple. This would be in areas that might get covered with a label. One should consider putting a finish on a plastic part when you need a first-class part and cost is not critical.

QUALITY CONTROL

Quality control is a subject everybody talks about but few people understand. It does not have to mean white lab jackets and elaborate laboratories. For the project manager it should mean an understanding with the supplier. What is being purchased from that supplier? More important, what condition must it be in before you will accept it as salable merchandise? Simple questions but ones that rarely get answered to everybody's satisfaction.

One should use every tool available to define just what it is that you expect to purchase. This includes engineering drawings, written descriptions, and even signed-off samples of what you demand. When you are dealing with a painted or decorated part, the signed-off sample becomes necessary. It is almost impossible to define clearly how a part should look on a piece of paper.

Dimensional tolerances are also important. Molding conditions change, as do materials from lot to lot. If you need a tight tolerance on a part, spell it out. On the other hand, if you do not need tight tolerances, they should not be put on a print just because they look good. Unnecessary tolerances cost you money and waste valuable time.

The most important thing is to have a good working relationship with the vendor. Tell the vendor why you need close tolerances and work with him to reach an acceptable solution to the problem. There are many tricks to the trade, and your vendor, if he knows what you want, can usually find a way to meet your criteria.

Such things as shrink fixtures, use of blowing agents, and general modifications of the molding conditions are all things that can alter a part.

There is a trend these days to require suppliers to furnish parts to their customers that meet definite quality control requirements. Doing away with incoming inspection reduces customer costs and saves time. However, it is obviously risky if you are to assume that all parts meet your requirements and put them into inventory without inspecting them.

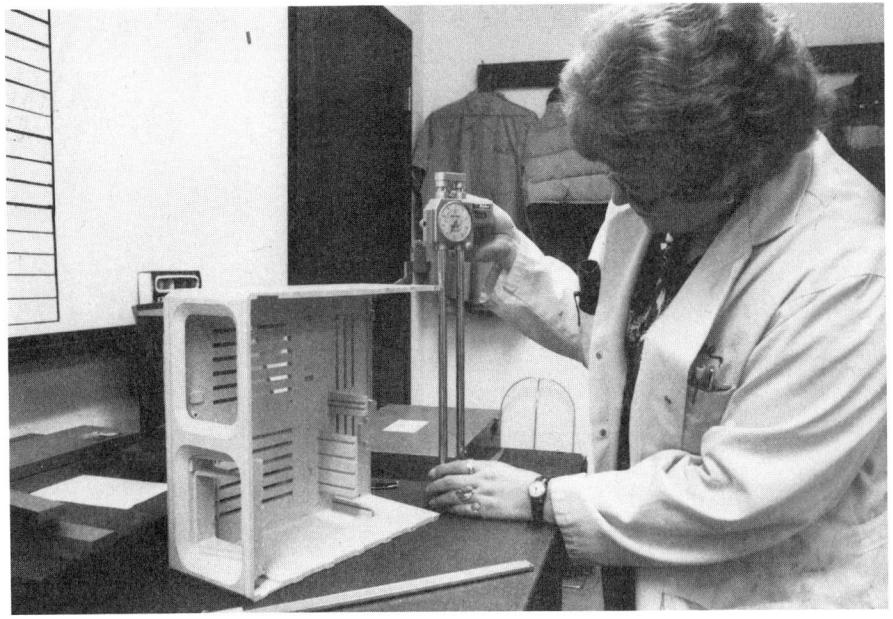

Figure 19
Precise measuring equipment is used to control dimensions of a molded plastic part. (Photo courtesy of Production Plastics.)

Here again, a close relationship between vendor and customer is necessary.

When one chooses a vendor, it is important to determine what kind of quality control facilities they have. Procedures for inspecting parts coming from the line are important and it is necessary that the testing equipment be available to do an adequate job. The vendor should also have counting and weighing scales. It is costly to receive shipments that are miscounted, and the necessary paperwork to correct these mistakes is a burden.

Again, communications are the important thing. If your vendor is continually trying just to get by with shipments that are not quite what you want to buy, you are in for problems. No one likes to take products back after they have been shipped. Talking these problems out and clearly establishing quality control requirements you can both live with will go a long way toward establishing a good working relationship.

5

Plastic Processing

DETERMINING WHAT PROCESS TO USE

Choosing what plastic process to use when producing a plastic application is often very difficult and can be expensive and time consuming if you pick the wrong one. The first question to ask is what volume per year the application is going to utilize. This can be a difficult question to answer, but to make a decision on what process to use, you need to come up with a realistic value. If the answer is 1000 parts or better per year, you can assume that an automated process should be used. Fewer than 1000 parts and you should consider fabricating the part from blocks of material, either metal or plastic, depending on end use.

The next question to ask is how large the single largest part you plan to produce is going to be. If the part weighs more than 5 lb, you can assume that finding an injection molding machine large enough to run this size of part is going to be difficult and that the tooling to produce it will be very expensive. A process such as structural foam, vacuum forming, or perhaps roto-molding would be more appropriate.

The final question on process selection is whether you require detailed wall sections and other refinements, such as bosses, side ribs, holes, and other detail. The injection molding process with its matched metal tools is about the only process that will provide this. However, sheet molding, compression molding, and vacuum forming have made significant strides in this area.

The vacuum forming process is a good way to make large parts, but added ribs, bosses, and holes raise the cost of each part due to post-forming fabrication. If the part you have in mind is hollow in nature, blow molding or roto-casting should be considered. Close-tolerance parts require close control on processing equipment and usually are produced best on injection molding machinery. Large flat surface are expensive to produce by injection molding and lend themselves to vacuum forming from extruded sheet. If the volume is low, you may wish to consider fabricating from sheet stock.

The smaller and more intricate the part, the more likely it is to be more economically produced by injection molding, possibly with multiple cavities. This depends on the volumes required. Multiple cavities are also used on vacuum formers, and a large number of parts can be produced in this way. Food containers produced from foamed polystyrene are an example of this.

When a set of parts is required to produce an assembly, family molds both for injection molding or structural foam may be the answer. There are several problems with the family mold approach, such as different part weights and different product ratio requirements, but the technique is still used. When a part has a defined profile and requires multiple lengths, the obvious method to consider is extrusion. Fairly decent volume is required for this process, but it is an economical way to produce parts that have this configuration.

Nearly any part can be injection molded; however, undercuts and complex configurations not in the direct line of mold opening are expensive, in both tooling costs and molding time. Some materials are more forgiving than others and can be molded with partial undercuts. One should discuss such problems with molder and your material supplier.

Various miscellaneous processing techniques are being used today to provide parts of unusual material combinations. Such methods as co-injection, twin sheet vacuum forming, and other types of insert molding have made it possible to combine the properties of more than one material into a single plastic part. Often, a molder will specialize in one or more of these techniques. It is best to check with your material supplier to find a molder who specializes in a particular process.

Sometimes there will be more than one choice when selecting a process. Nearly all thermoplastic materials can be processed on any thermoplastic molding equipment. Some thermoset materials need more specific machinery for production. An example is that of the urethanes, which are normally produced on a multicomponent mixing system and thus require a casting process.

When there is a choice of processes, it is important to investigate all the parameters of each. This includes tooling cost, tolerance requirements, raw materials available, and part costs. When one is selecting a process, the method of assembly should be taken into consideration. It is important to know how you are going to put parts together during assembly. You may want to consider inserting metal inserts for joining two parts. These could be molded or put in as a post-molding operation. It is possible that you do not want to use fasteners at all, possibly using a snap fit to hold two parts together. Adhesive bonding may be a third choice, or possibly ultrasonic bonding should be considered. Again all possible methods should be investigated, looking at the insert cost, holding force needed, length of time needed to bond or insert, and added cost of handling the part after it is molded. Whether the part is exposed to oxidation should be considered as well, and the type of insert selected accordingly.

An example of this selection process is that of an electrical connector being purchased by a small electrical manufacturer. They were buying a 12-in. connector from an outside source at a cost of $17. The part was produced from a thermoset material using a compression molding process and had six brass inserts molded into it to fasten electrical wires. The manufacturer wanted to investigate the possibility of purchasing the part for less while maintaining the same physical and electrical characteristics.

The only way to produce the part was to injection mold the part using a thermoplastic. Although the volumes were low (2000 per year), the cost of an aluminum tool was investigated. It was decided to go with a chemical blowing agent and polycarbonate to get the required electrical properties and to take care of the thick sections in the part. The brass inserts were to be put in with a ultrasonic machine as a post-molding operation.

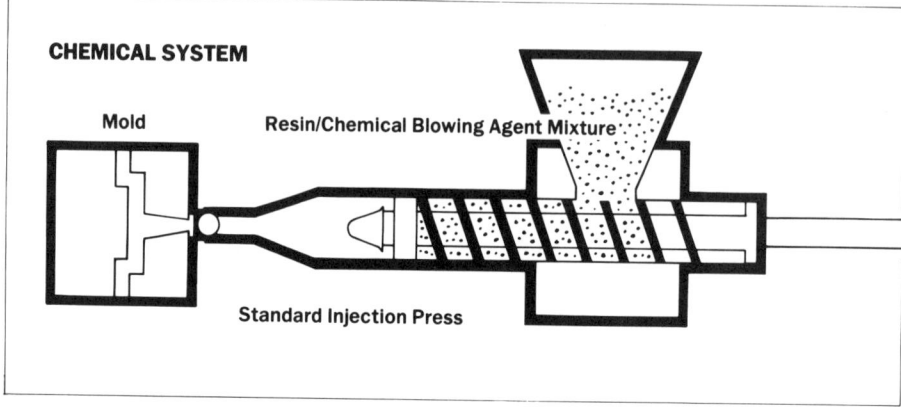

CHEMICAL SYSTEM

Mold   Resin/Chemical Blowing Agent Mixture

Standard Injection Press

Figure 20

The use of chemical blowing agents has made it possible to process structural foam on conventional injection machines.

The mold was built and first articles molded. It was determined that it was less expensive to load the inserts into the mold and mold the material around them. The part came in at a price of $5 per part with a mold cost of $5000. The result was a lower-cost article with equal if not better physical properties and savings large enough to pay off the tool within one year.

The important point here is to make sure you investigate all processes that seem to have merit. There may be several ways to produce an item, but once you have reviewed all possible methods, one or possibly two methods will stick out and should be evaluated thoroughly.

The many processes available to mold or form a plastic product vary in numerous ways. Each has a distinct means of taking the raw material and shaping it into a useful object. In this chapter we explore the various processes and the products that can be produced from them.

BLOW MOLDING*

Blow molding is an efficient molding technique used to produce containers and bottles from thermoplastic polymers. In this process a tube

---

*Material in this section has been taken from E.L. Dolliff, "Blow Molding," *Modern Plastics Encyclopedia*, McGraw-Hill Book Co., New York, 1981–1982, pp. 234–241.

or *parisan*, as it is called, is extruded. This is immediately captured between halves of a metal mold that pinches off the parisan at the top and bottom. Air pressure is inserted into the center of the parisan, which blows the plastic against the sides of the mold. The mold is then cooled and the container removed. Since the mold must be brought to the parisan, this can be done in one of two ways: by shuttle, which moves back and forth between two stations, or by a rotary, which moves between several stations in a circular pattern. The second way is the most efficient but does not lend itself to short runs, due to the excessive setup time required.

Other techniques of blow molding exist and usually consist of injection molding a tube-shaped parisan with the threaded top of the container molded in place. This parisan is then moved to where it can be heated and put between the halves of a metal mold. At this time air pressure is introduced into the center of the parisan, blowing the molten plastic out against the sides of the mold, forming a bottle with the threaded top attached. Molds for this process are usually made of cast aluminum.

Applications. The areas of application for blow molding are usually found in the container field. Since the process produces a hollow-skinned plastic item, the products produced are usually such things as bleach bottles, gasoline containers, and more recently, gas tanks for vehicles. Almost any container can be produced, some with handles, others with molded-in spouts. One of the most recent has been the blow molding of motor oil containers to replace the standard quart can. Since not much pressure is involved in the process, very large containers, such as 55-gallon barrels, have been produced.

COMPOSITES*

Many changes have taken place in the field of composites. In this area of technology, polymer engineers have been able to build a variety of new products by embedding load-bearing fibers in a plastic matrix. Aircraft and automotive designers have taken a new look at this development with an eye toward building all types of vehicle structures.

---

*Material in this section has been taken from Donald Dreger, "The Challenge of Manufacturing Composites," *Machine Design*, 1987, pp. 92–97.

This new material technology also demands improved manufacturing technology. It is here that much improvement is needed, for the industry suffers from too labor intensive a manufacturing process. Also, this labor must be skilled to perform the intricate techniques needed to produce a quality product. Because of this, operations are generally costly and time consuming.

Fibers used are generally high-cost carbon and boron, and the principal matrix material has generally been epoxy for aircraft and polyester of other applications. However, new thermoplastic resins have been developed with improved high-temperature properties. Compared to thermosets, these thermoplastics can be cured in shorter periods of time, saving much processing time and cutting costs significantly. Another benefit is the indefinite shelf life offered by thermoplastics and also the fact that these materials may be reground and reused if necessary.

Thermoplastic polymers used for composites include polyarylene ketone, polyether ether ketone, polyphenylene sulfide, polyamide imide, polysulfones, and liquid-crystal polyester. All have processing temperatures in the range 624 to 788 °F. Concern with these materials lies in the creep and fatigue resistance area, and much work is being done to improve on these characteristics.

Not one, but several processing techniques are used to produce composties. These include hand layup, spray-up, vacuum-bag molding, pressure-bag molding, autoclave molding, pultrusion, resin transfer, braiding, injection molding, and expanded polystyrene foam molding.

Hand layup is the simplest and oldest techniques, but production is low in volume and very labor intensive. Glass or other reinforcing matt or woven roving is positioned manually in the open mold and resin is poured, brushed, or sprayed over and into the glass. Entrapped air is removed manually with a squeegee or roller. Room-temperature curing polyesters and epoxies are the most commonly used resins. Curing is initiated by a catalyst or accelerator in the resin system, which hardens without the application of internal heat. For a high-quality part surface, a pigmented-gel coat is first sprayed on the mold surface.

Spray-up is an open-mold method that can produce complex parts more economically than hand layup. Reinforcing fiber and catalyzed resin, and in some cases, filler, are deposited in the mold from a combination chopper/spray gun. As in hand layup, gel coats can be used to produce a smooth, pigmented surface.

Vacuum-bag molding is a refinement of hand layup and uses a vacuum to eliminate entrapped air and excess resin. A vacuum is drawn on a film bag, putting pressure on the composite. It is then cured at oven temperature. This vacuum method provides higher reinforcement concentration and better adhesion between layers.

Pressure-bag molding is similar to the vacuum bag method except that air pressure, usually at 30 to 59 psi, is applied to a rubber bag or sheet laid on the composite. This forces out the entrapped air and excess resin. Pressurized steam may be used to accelerate the cure. Only female molds are used with this method.

The *autoclave method* is a modification of both the pressure-bag and vacuum-bag molding techniques. An autoclave, a heated pressure vessel, is used into which the bagged layup on its mold is taken for the cure cycle. Cures are faster and the method produces denser, void-free moldings because heats and pressures are higher. The process also works better with higher-temperature matrix resins which have higher properties than conventional resins. The use of an autoclave limits the part size, however.

The *pultrusion method* of processing composites is a continuous technique used for producing constant section shapes, such as ladder rails, beams, and pipe. The reinforcement, usually a combination of continuous strand roving, woven roving, surface mat, or veil, is pulled through a thermoset resin bath to wet out the fibers. Excess resin is removed and the saturated reinforcement is pulled through a die, then through a heated jacket to cure the resin. The finished shape is cut to lengths by a traveling cutoff saw. High-strength parts are possible because reinforcement can be as high as 75%.

Resin transfer molding is a low-pressure process for moderate-volume production. Mat and woven reinforcement are laid up dry in the bottom of a matched surface mold. The mold is closed and clamped and a low-viscosity catalyzed resin is pumped in, displacing the air and venting it to the mold edges. The resin is allowed to cure and the part is then removed. The advantages over other methods are that a uniform thickness is possible and both sides of the mold can be coated with a gel system to provide smooth surfaces. This method is faster than compression molding.

Braiding is a rapid reinforcement method that produces strong interwoven tubular or flat structure from glass, carbon, or amid yarns. The braids are laid up (usually over a mandrel) wet or from prepegs and autoclaved, or laid up dry and finished by resin transfer molding.

Injection molding is the most widely used process for high-volume production of thermoplastic resin parts.* With modifications, the process can be used with thermosets as well. Pellets of resin containing fiber reinforcement are fed into a hopper and then into a heated barrel containing a rotating screw that mixes and heats the material. The heated resin is then forced to high pressures through sprues and runners into a matched metal mold. Molding cycles are rapid and parts can be very precise as well as complex.

Expanded polystyrene foam molding utilizes beads of styrene fused together by means of steam heat and formed into a molded shape by means of a cast metal mold. The equipment used consists of a low-pressure hydraulic press and a conveying system that inserts the preexpanded styrene beads into the mold. Here they are exposed to steam that fuses the beads while expanding them further so they fill the mold. The parts are then cooled in the mold by means of cooling water and are then removed from the mold and allowed to dry off before packaging. The density can be controlled by controlling the density of the preexpanded bead before it is introduced into the machine.

Molds are usually of cast metal (aluminum) and are built to withstand the low pressures that are present in the system. Air valves for steam entry are built into the sides of the cavities, and molds have water jacketing for cooling. Applications include ice coolers, insulative liners for water jugs, packaging products of all kinds, and void fillers.

EXTRUSION[†]

Extrusion is a plastic process in which a plastic resin is heated to a continuous melt and forced through a die to form a profile shape. The melted material must then be cooled back to its solid state as it is held in the desired shape. A product produced in this manner is in a continuous shape which must be cut into the lengths desired. Products produced in this manner include pipe of many different diameters, house siding,

---

*Material in this paragraph is from *Machine Design*, Oct. 22, 1987.
[†]Material in this section has been taken from J.A. Gibbons, Dan Grynberg, and Hassan Helmy, "Extrusion," *Modern Plastics Encyclopedia*, McGraw-Hill Book Co., New York, 1986–1987, pp. 219–234.

Figure 21
Typical extruder line producing thousands of feet of plastic extrusion. (Photo courtesy of Poliwood Molding Ltd. and Prodex.)

auto trim, sheet stock of all thicknesses, coatings such as wire insulation and pipe coatings, and many other useful shapes. It is also used to produce long, thin strands of plastic which are chopped to form beadlike pellets used in other plastic processes.

The equipment used for this process is usually a single screw extruder with the ability to process the widest range of polymers in the most economical manner. These extruders range in size from ½ to 12 in. and vary in length according to use.

The entire system consists of a hopper from which either granules or powder is fed into the feed section of a barrel, where it comes into contact with a screw. The screw is driven by a motor at rates of 20 to 200 rpm. The screw is rotated in a steel barrel having a very hard, wear-resistant liner. The barrel is electrically heated to aid in the melting of the material. Cooling jackets are also used to control the temperature profile of the melt passing through the barrel. Temperatures in the various barrel zones are set according to the material, screw design, and processing goals. The zone temperatures vary widely depending on the material used or the product being made.

The screw is the main part of the extrusion system, and its design is of critical importance when selecting the equipment for a specific job. Its main function is to propel the material through the barrel and to be sure that heat is distributed through the melt. It also assures that the melt is uniformly mixed prior to forcing it through the die and out into the cooling and sizing equipment.

The melt must be shaped and cooled by sizing and cooling equipment while it is returning to its solid state. The method of creating the end product varies depending on the shape desired. Sheet products are cooled and sized on highly polished and liquid-cooled rolls and then wrapped in continuous coils or cut into specific sizes.

Pipe or tubing is usually cooled through open water troughs or pulled through a vacuum sizing tank where the melt is held to a sizing sleeve. Blown film normally is extruded vertically into a tubular air-blown shape and up through a tower, where it is collapsed, slit, and wound up in rolls.

Fiber is extruded out of a die with many small orifices that allow thin strands to be drawn down to specific thread or filament sizes.

Custom profiles are commonly made of material that has a high melt viscosity so that the profiles will hold their shape while they cool. They are usually made at low processing speeds and are cooled by forced air, in water troughs, or by a water spray method. Various sizing fixtures are often used to hold the extrudate while it is being pulled through the cooling system.

Becoming more important is *coextrusion*; here multi-layers of different materials are extruded on each other to form a product of different properties. This require multiple extruders and a specialized die system to bring the layers together before putting the combination product into common sizing and cooling equipment.

The tooling or dies needed for extrusion are relatively economical, but complex shapes require considerable knowledge and experience for satisfactory results. If you need a plastic part with a continuous profile, it is worthwhile to investigate extrusion first. Nearly all of the various thermoplastics available can be extruded, and tooling is much less than for injection molding or some of the other processes.

INJECTION MOLDING

Injection molding is one of the more common plastic processes, and a great number of products are produced using this efficient and reliable method of making plastic items. The principal advantage of this type of molding is that you can control the thickness of the skin or wall of the part. By designing the mold correctly, a part that meets all the dimensions and performs functionally can be obtained. The biggest disadvantage of the injection molding process is the high cost of tooling. When you buy an injection mold, you are buying an expensive piece of machinery—a piece of machinery that will possibly produce thousands of good parts and require little maintenance.

First, let's describe the process. Plastic in the form of small pellets, sometimes called resin, is fed into a hopper at the rear of an injection molding machine. From this hopper this material is gravity-fed into the rear of the barrel, where it is passed through the screw by means of a screw device. In the barrel the material is heated both by the mechanical movement of the screw and by electrical heat applied by heater bands positioned along the barrel length. Under heat and pressure, the material, mainly thermoplastic in nature, is melted to a liquid form, which is then forced from the front of the barrel, through a nozzle, and into the mold. Here it is held while it cools down, usually helped by a water-filled cooling line that runs through the mold. When the cooling process is complete, the mold is opened by a hydraulic press attached to the mold and the part is ejected from the cavity. The process is now ready to start over. The time needed for this filling and cooling process to take

Figure 22
A Large 1500-ton clamp injection molding machine manufactured by Cincinnati Milicron. The unit can shoot 50 lb of plastic at once by making use of a second plasticating unit mounted atop the primary barrel. (Photo courtesy of Cincinnati Milicron.)

place is called the *molding cycle*. It can be affected in many ways by such things as part design, mold design, material, and machine function.

Troubleshooting Injection-Molded Parts.   Often when you receive the first sample article you may want to know what is causing the problem. You may want to be able to discuss the problems with your molder, and it is helpful to know enough about the molding process to discuss it intelligently. Some problems and their possible causes are discussed

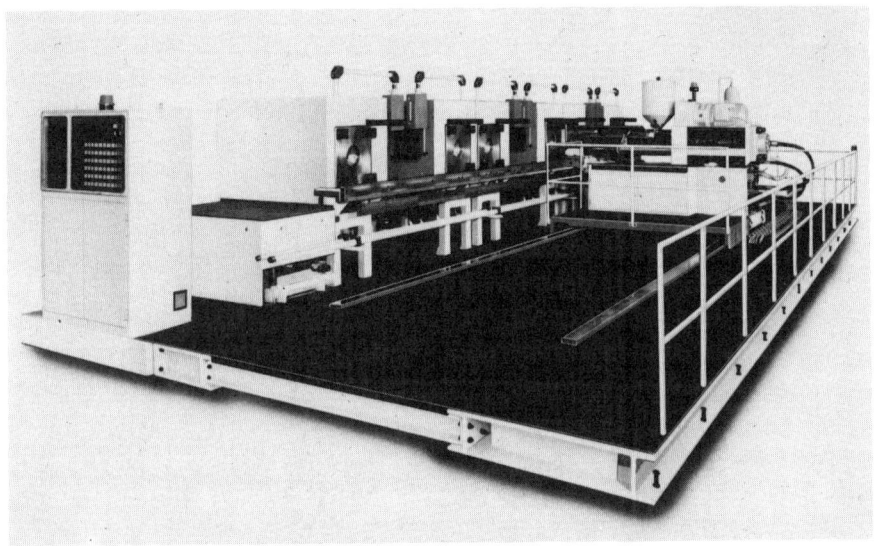

Figure 23
German molding machine mounted on a track that allows the injection unit to feed several mold stations consecutively and increase the productivity of the system. (Photo courtesy of SPI.)

below. As the causes described may not be the only ones possible, you should seek the advice of an experienced molder.

Problem 1: Bubbles or voids in the part
Possible cause: These bubbles may be caused by a number of things. They may be caused by too low an injection pressure, entrapped air that requires more venting, too low a mold temperature, or too low a clamp pressure. They might also be caused by material flowing from a thin section to a thick section.

Problem 2: Delamination—two obvious layers of material showing up on the part
Possible cause: This can be caused by incorrect mixing of regrind or the possibility that the machine was not properly purged, causing the introduction of two or more incompatible materials.

Problem 3: Sunken areas opposite a perpendicular rib
Possible cause: This could be caused by too low an injection pressure, by improper part design (rib thickness should be 70 to 80% of the wall thickness they run into), by incorrect gate design, or by too little injection pressure hold time.

Problem 4: Silver streaking or splay on part surface
Possible cause: This is usually caused by inadequate drying of material, possible water leaks from the cooling system directly into the cavity, sharp corners around the gate area, too small a gate, or too high a nozzle temperature.

Problem 5: Appearance of visible knit lines
Possible cause: Possible causes are having the mold too cold, injection speed too slow, insufficient venting, or too great a distance from the gate.

Problem 6: Excessive flash around parting lines
Possible cause: This can be caused by having too low a clamping pressure, too high an injection pressure, too-large vents, or too fast an injection speed.

Problem 7: Low-gloss-level areas on the part.
Possible cause: This is caused by too cold a mold temperature or by having too low an injection pressure or to slow an injection speed.

Problem 8: Warpage and distorted part configuration
Possible cause: The causes for this are possible uneven part ejection; cores and cavities at different mold temperatures; cooling time too short, or injection pressures too low.

Most of these problems can be solved by changing the machine process conditions. One should immediately be aware of problems that can cause damage to a tool, such as flash. One does not have to accept poorly molded parts from a vendor. Sometimes altering a mold will cause added expense, but usually there is a way to cure any problem.

Estimating Cost of an Injection-Molded Part. It is often necessary to have a rough estimate of the cost of an injection-molded part before you do a lot of detailed design work. To do this you must at least have a rough sketch of the part, showing width, length, depth, and the thickest wall section of the proposed part. With the width, length, and depth, you can calculate the square surface area. Multiplying this by the average wall thickness will give you a good estimate of the cubic inches of material you will use.

The next step is to find the weight in pounds that your part will use. This is done by multiplying the calculated cubic inches by a conversion factor, 0.036, and then multiplying this sum by the specific gravity of the material you plan to use. If you have not decided on a material at this time, you can use an average value of 1.5. This will give you too light a weight if you plan to use a glass-filled material, but it should be usable for a cost estimate.

To find the material cost of the item, you multiply the weight of the part in pounds by the cost per pound of the material you plan to use. If you have not determined what material you plan to use, you can get an estimate by using the following table:

Polymer	Cost per pound
Commodity resins	
Styrene, polyethylene, polypropylene	$ 1.00
Common engineering resins	
ABS, SAN, polycarbonate	2.50
High-heat engineering resins	
PEEK, PEI, PAS	15.00

The next step is to determine the cost of molding the part. If you convert the weight of the part to ounces, you can determine the size of machine required by using Table 1. Keep in mind that the molder may want to put the part on a different-size machine for some reason. Remember to multiply the weight of the part by the number of cavities if you plan to use multiple cavities.

Table 1 Determining Machine Size and Cost Given Part Weight

Part weight (oz)	Machine size (tons)	Average hourly machine rate
less than 5	50–75	$ 25
5–10	100	35
10–20	200	50
20–40	300	70
40–60	400	80
60–90	500	90
90–150	700	100
150–250	1000	120

To determine the cycle time you must go back to your part and select the thickest wall section found. Use Table 2 to find the cycle time that would be required.

From these values we can establish what the part will cost to manufacture. To this value we add the material cost, a scrap and reject factor, and an average setup charge. The sum of these should provide an estimated cost to produce the article. A scrap factor of 20% is fairly common in the industry. To obtain the setup charge, use an average of $150 per 1000 parts, or an additional cost of $0.15 per part. For a 12 × 12 in. flat part ¼ in. thick, you would have 12 × 12 × 0.25 = 36 in.³ × 0.036 = 1.296 × 1.5 = 1.94 lb or 31 oz of part. The material cost for polycarbonate would be $3.00 × 1.94 = $5.82.

Molding cost would be figured using a 31-oz machine or 300 tons at $70 per hour. For a 0.250-in. wall section the cycle time would be 72 s or 50 shots per hour. This would calculate to a machine time cost of $1.40. Add this to the material cost of $5.82 and you get a cost of $7.22. Add a scrap factor of 20% of $1.44 and an additional $0.15 for setup and you get an estimated cost of $8.81. One should keep in mind that these costs are only estimates—you need to have a qualified molder estimate the cost at the appropriate time.

Table 2 Cycle Time Determination

Thickest wall section of part (in.)	Cycle time (sec)	Shots per hour
0.039	13	277
0.049	15	240
0.099	25	144
0.124	32	112
0.148	38	95
0.174	48	75
0.195	60	60

ROTATIONAL MOLDING*

Essentially, rotational molding is a process for producing seamless hollow plastic parts suitable for use as containers and other parts. The process consists of putting either powdered or liquid polymer into a closed mold, heating the mold to the melting temperature of the polymer, and rotating the mold about two perpendicular axes simultaneously. This forms a layer of polymer with uniform thickness. The mold is then moved to a cooling station, where it is cooled down by the use of forced air and/or water spray. From there it goes to a third station for demolding. The equipment usually consists of a three-armed rotational molding unit where one arm is always located at one of the three stations. One-armed units are also used but are obviously not as efficient. Materials used are generally thermoplastics, but some thermoset materials and cross-linked polyethylene have been tried with some success.

Tooling for this process involves low-pressure inexpensive molds and is usually made in cast aluminum. Electroless nickel molds have been used but are more expensive. For prototyping, molds have been constructed using glass-filled epoxy, but because of the temperatures used in the process, are limited.

---

*Material in this section has been taken from R. L. Fair, "Rotational Molding," *Modern Plastics Encyclopedia*, McGraw-Hill Book Co., New York, 1986–1987, pp. 400–402.

Figure 24

Roto-molding machine that utilizes a rotating arm mechanism to put the mold first into an oven and later into a cooling chamber, where the molten plastic can harden prior to removing it from the mold. (Photo courtesy of McNeil Akron).

Applications for rotational molded products are of great variety and include jugs and containers of all types: refuse carts, toys, shipping containers, boats and canoes, swimming pools, and hundreds of similar products.

REACTION INJECTION MOLDING*

Reaction injection molding is a process whereby two liquid streams of chemicals are metered and mixed in a mixing head and dispensed into a mold, where they polymerize and form into the shape of the mold. Normally, urethane polymers are used, but recently other types of polymers have been used. The finished product can be in the form of

---

*Material in this section has been taken from *Rim Reaction Injection Molding*, RM-8206 (10)L, Mobay Chemical.

foam or a solid, and the hardness can be adjusted to provide a very hard material or a fairly flexible one. The great feature of urethanes is that they can be formulated to almost any set of properties that you would want.

The form of mold-release agent used is important and the parts are often labor intensive, due to required cleanup of flash and excess mold release. Sometimes it is difficult to get consistent parts over a large-volume run. Other problems can be encountered due to improper metering of ingredients, and high humidity can often cause variation in the product.

Most rigid parts get painted before they are used, and inserting metal or other materials is common. RIM equipment is highly modular. The dispensing system can serve one mold or several. Add-on units can be incorporated for such additives as flame retardants, colorants, blowing agents, or fillers. Additionally, the temperature of the liquid reactants, flow rates, and densities can all be monitored and controlled with microprocessors.

Although RIM polyurethanes have made the greatest inroads in automotive applications, more designers and engineers are turning to RIM urethanes for industrial and consumer products. Moreover, even faster RIM urethane systems have been developed, and equipment and techniques have been refined to the point where RIM has firmly established itself as a leading forming process. Mold construction is fairly inexpensive and cast metals are usually used. Often, heating jackets are required around the tooling, and ejection systems can be simple in nature.

The molding possibilities of RIM are great. Complex shapes, variable wall thickness, soft lines, inserts, encapsulations, texturing, ribbing, and bosses are among the design possibilities. In addition to thick or thin parts, RIM urethanes are suitable for molding parts large and small. They can range from 100 lb down to less than 1 lb. When it comes to aesthetics, RIM urethanes can duplicate the mold surface faithfully to product textures, wood grain, and even smooth, glasslike surfaces.

Other developments include the introduction of glass fiber and flake reinforcement to RIM systems. Processing structural RIM systems combined with the chemistries of traditional urethane allow for the production of composite parts with complex shapes and excellent and consistent physical properties. In simple high-volume applications, steel and other materials could still be economically competitive. However, the trend to smaller volume production and improvement in RIM systems, such as internal mold release and better, more durable surface quality, promise a continual and growing interest in RIM.

STRUCTURAL FOAM MOLDING*

Structural foam is a molding process very similar to injection molding. It has often been defined as low-pressure injection molding and varies only in that a blowing agent is added to the melt prior to molding. This forms a product that has a cellular core and integral skins. Because the mold is not packed with material by the molding machine, a structural foam part has very little stress built into it, such as with standard injection molding.

Two basic types of machinery are used in producing structural foam. First are the specially built machines designed to produce large parts of structural foam, up to 120 lb. This equipment consists of a large hydraulic press attached to a large extruder into which thermoplastic resin is fed. In the extruder the polymer is heated by a combination of electrical heaters and by shear energy caused by the turning of the screw in the barrel. Once in the melt stage, the material is mixed with the blowing agent, usually nitrogen gas.

This mixture is then introduced through a manifold system into the mold, which is usually made of machined aluminum. The material surges into the mold much like shaving cream. As it hits the cold mold wall, the first layer of cells break down and form a skin. The rest of the foaming material is left in the center of the part and forms a cellular core as it cools down. Because the mold cavity is filled by the foaming action, not by the hydraulic action of the machine, the part develops relatively few stresses. This provides a plastic part that is less susceptible to breakage and more resistant to chemical exposure or UV attack. Because these machines normally use a multinozzle manifold system, they can deliver the melt directly into one large part or a group of several smaller parts. This makes it economically feasible to produce a total assembly of parts in one shot. An example would be a complete computer

---

*Material in this section has been taken from Bruce Wendle, *Structural Foam: A Purchasing and Design Guide*, Marcel Dekker, Inc., New York, 1985.

Figure 25
Large 500-ton structural foam molding machine capable of shooting over 100 lb of plastic in a single shot. (Photo courtesy of Springfield, Inc.)

housing where the hood, base, bezel, and keyboard could be produced in one cycle of the machine.

The second type of equipment used to produce structural foam is a standard injection machine. By using chemical blowing agents mixed with resin, an injection molding machine can be used to melt and mix the blowing agent and dispense the mixture into a low-pressure mold in much the same way as is done in the large structural foam machines. Often, the shot size available on the machine is too little and the clamp pressure available is more than necessary, but one can still make excellent foam products.

In both cases the surface of the foam parts have what is known as swirl on them and may not be acceptable unless they are covered with a coating of paint. Much work has been done to eliminate this swirled surface, and there are several methods available that improve on this

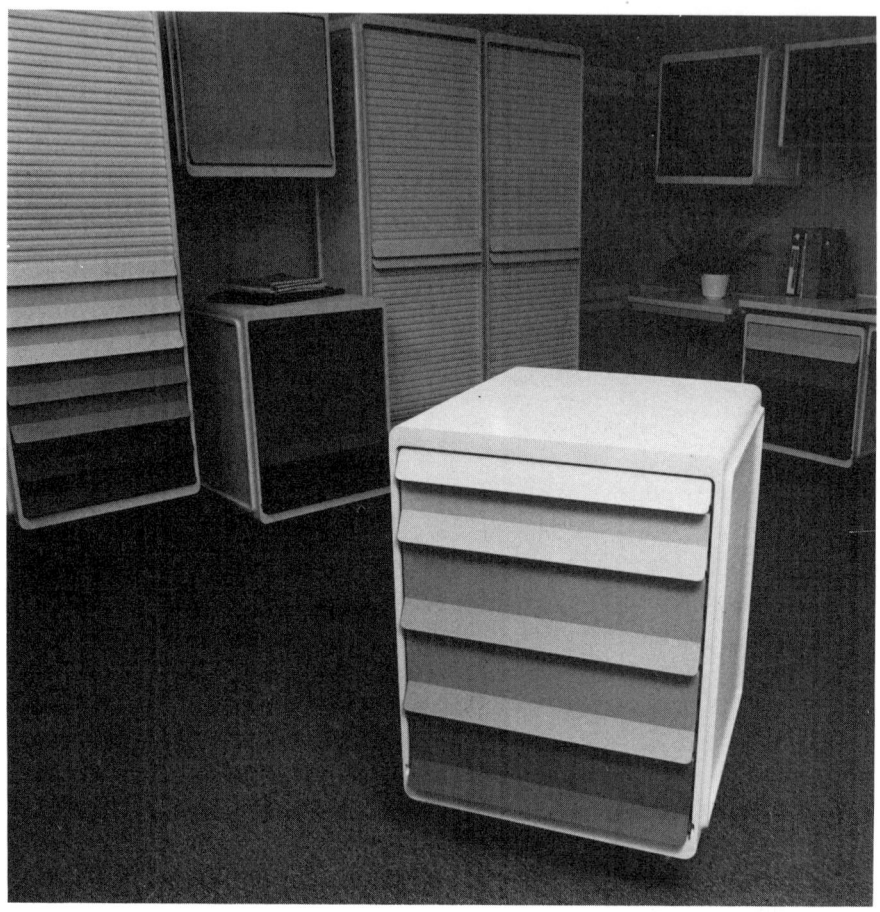

Figure 26

Cabinet system produced from structural foam. The total system features cabinets of various sizes and has drawers and replacable drawer fronts. It is molded on a multinozzle structural foam machine. (Photo courtesy of Zytec, Inc.)

imperfection. One method consists of sealing the mold cavity and putting a positive back pressure of some gas against the incoming gas plastic melt mixture. This keeps the gas from coming out of solution until the skin has been formed and leaves a smooth surface on the foamed part. This *gas counter-pressure system* has found some application.

Many applications have been found in which the swirl surface is not objectionable, while a good many more utilize a coating of urethane to cover it up. This really provides an excellent system in which the paint acts as a scratch-proof colored barrier that adds greatly to the function of the part.

The tooling normally used with structural foam is often the system's biggest advantage. Made from machined aluminum and designed for the system's low pressures, the cost is usually 40% less than that of the high-pressure steel tooling needed for standard injection molding.

Most available thermoplastics have been foamed with success, including the elastomers. High-impact styrene is especially well suited to structural foam and provides a tough, impact-resistant material suitable for many applications. High-density polyethylene, with its resistance to moisture and general all-around toughness, also finds many applications in foam form. Some of the new high-heat materials will doubtless be used in aircraft and similar applications as foam.

We should note that there are certain limitations with structural foam. By replacing the center of a structure with a cellular core, the physical properties of that structure are reduced. The cycle time to foam a part is also increased because the foamed material does not cool as rapidly as the solid material. This often results in an increase in cost of the foamed part over the solid, molded part. However, the ability to produce a large molded part, where you have complete control over the wall thickness, can only be accomplished in structural foam.

VACUUM FORMING*

Vacuum forming is a process whereby a sheet of thermoplastic is heated to an optimum forming temperature and then by means of a vacuum and sometimes air pressure, formed into a shape. It is a process that

---

*Material in this section has been taken from *Thermoforming to Optimize Performance*, Dow Chemical Co., Midland, MI, 1986.

Figure 27
Horizontal structural foam molding press, which was the first type of press used with the process. (Photo courtesy of SPI).

Figure 28
Large bins produced as a one-shot part molded in high-density polyethylene
structural foam. (Photo courtesy of Zytec, Inc.)

has gone through many changes since its initial commercial development in 1950.

The introduction of new polymers has kept pace with machine development and this has helped its growth. Present applications include automotive interior trim on door panels, instrument panels, headliners, floor mats, heater ducts, wiring harnesses, door pillar posts, gas tanks, fenders and fender liners, hub caps, bucket seats, and a variety of other parts.

Originally straightforward, simple machines, vacuum formers have developed into sophisticated solid-state-controlled equipment. This change has been brought about by a demand for more complex parts, faster cycles, deeper draws, and development of both rigid and flexible new materials.

The simplest thermoforming machines consist of a forming station, which is generally used for loading a plastic sheet into the clamping frame before forming and unloading it after forming. Next to the forming station is the heating station, which heats the sheet to the forming temperature (see Table 3). The operation starts with loading the sheet into the clamping frame, then moving it into the heating station, where it remains until it reaches the optimum forming temperature. It is then moved back into the forming station, where by moving the mold or the sheet, or both, the edges are sealed around the sheet, and the air between the sheet and the tool is exhausted by means of a vacuum. This forces the heated sheet to conform to the shape of the mold. It remains there until it is cooled to a solid state and can be removed from the clamping frame in its new configuration.

Nearly all existing vacuum-forming equipment evolved from this basic concept. Changes were brought about by the need to accomplish various changes in the process. These included such things as a need to heat the sheet faster, cool the formed part faster, assist in forming the sheet through the use of plugs, and control the temperatures of the sheet more closely. Also, automatic feed and unloading were needed, as was a way to trim the parts more accurately while they were in the mold.

One way of doing much of this was use of a rotary thermoforming machine. This machine was developed primarily for handling large rigid sheets faster than a single table machine. Normally, such a machine consists of a large rotating wheel that carries three or more clamping frames through the various stations. Thus a sheet can be loaded into

Table 3 Thermoforming Processing Temperature Ranges

Material	Normal processing temperature	Setup temperature
ABS	325	185
Acrylic	310	225
CAB	295	160
Polycarbonate	455	280
Low-density polyethylene	295	120
High-density polyethylene	375	180
Polyphenylene oxide	400	270
Polypropylene	400	250
PVC–acrylic	370	175

one of the frames at the first station, then rotated to the next station, which would be the heating station. It would then move to the third station, where it would be formed. After it was thermoformed; the part would either be removed from the clamping frame or rotated to the next station for removal. The material used could be up to ½ in. thick. It could be either rigid or flexible, manually or automatically loaded, heated from one side or from both, vacuum formed or pressure formed, and formed over a male tool or into a female tool.

Another type of machine was developed to rotate a number of molds around a forming station at the same time that a sheet is being rotated through the heating station. The two units rotate 180 degrees apart and two or three tools can be used at one time. This type of machine allows secondary operations, such as embossing, trimming, or edge turning, before part removal from the tool.

A third machine is the twin-sheet thermoformer. This machine is used to form two sheets simultaneously to make a variety of enclosed parts such as gas tanks, holding tanks, and house shutters. It is also possible to put inserts between the two halves before they are bonded together.

Another type of thermoforming machine is the in-line machine. It is used primarily to form material that can be fed into it from roll stock. This includes thinner gauges of rigid material (up to 0.090 in.), flexible materials, and thin-gauge foam sheet. The foam materials are used in egg cartons, plastic plates and cups, and similar packaging applications.

Although thermoforming has found many applications, it is a very labor intensive process, not so much in the actual forming of the parts as in the trimming and secondary work required on the parts after forming.

A wide variety of materials can be used to build tools for the forming process. Although aluminum is favored, such materials as wood, plaster of paris, polyester, epoxy, and polyurethane have all been used successfully. Steel is used where long life is required.

The economics make this molding process a good one to look into where volumes are small and you want an inexpensive mold. It also lends itself to parts that have a large square area. Because of the high cost of trimming and other secondary work, high-volume applications are difficult to justify unless the system can be highly automated.

6

Secondary Operations

As an engineer your job responsibility does not stop with simply molding the plastic part. Often, additional work is required to make the part usable in a production situation. Whether it is bonding to another part, machining the part, or painting, the techniques necessary to get the part production ready are important and you need to understand the procedures.

PAINTING*

Painting is often necessary to cover weld lines, knit lines, and other imperfections. It is often necessary to use a barrier coat on the raw part before applying the color coat. This is done to prevent the solvents in the outer coat from attacking the base polymer. The coating most often used on plastics for an industrial application is a urethane finish such as Sherwin-Williams' Polane. This provides a tough scratch resistant coating with excellent color control. Coatings with less scratch resistance,

---

*Some of the following material has been adapted from a booklet entitled, *Plastport to the Universe of Plastic Shapes for Machined Components*, published by ERTA Inc. of Exton, PA.

Figure 29
This unloading device makes it possible to keep product from an eight-cavity mold separated by cavity for quality control purposes. The machine is making thousands of 35-mm slide mounts. (Photo courtesy of Production Plastics.)

such as acrylics, are available. Epoxy coatings are also used. You should consult both the material supplier and the paint supplier before using a paint system on a plastic part. Incompatibility is often a big problem with painting systems, and blisters and bubbles in the paint surface can plague any plastic development.

Sometimes it is necessary to put a color coat over a precolored part to cover the weld lines and other imperfections. When you do this, keep in mind that if the undercolor is lighter than the paint coat, you may have problems with show-through when the top coat is scratched off. Often you have to sand the parts before painting to remove blemishes and possible sink marks caused by improper design or processing. This can be very expensive and also presents a problem for vendors not set up to handle sanding.

Quality control of painted parts is very necessary. Small bubbles may appear on the painted surface hours after a part comes off the paint line. This and color control put a high burden on the quality control system. The appearance of painted parts can be very subjective, and problems with painting are probably the leading cause of difficulties with a vendor supplying this service.

INSERTS

Such things as inserting metal inserts into a molded part to provide threaded holes are often less expensive than to have the insert molded into the part. Driving energy can be furnished by ultrasonics or by a heat gun, and the process takes very little time. The molder should be asked whether the insertion can be done at the machine. If the operator has enough time between cycles of the molding machine, this service can often be provided at no extra cost.

Along with threaded inserts, threaded studs are often inserted into the molded part by similar means. This often calls for a special horn to be provided, at an extra cost to the customer. Table 1 shows insert sizes and recommended hole sizes for specific inserts.

BONDING

Bonding two plastic parts together is another secondary operation often undertaken by the molder. This can be done either by ultrasonics or by use of an adhesive. Much work has been done on the development of new adhesives in recent years, and their application is not as difficult as it once was. Epoxies, urethanes, and cyclanurates have all found use in bonding two plastic parts or a plastic part and some other material, such as metal. It is still less expensive to try and mold the two parts as one rather than to bond two parts together, but often, this is not possible.

Ultrasonic bonding, on the other hand, can be used very efficiently. All that is required is that the parts are designed properly for bonding and that you have the right ultrasonic machine with a properly designed horn. The process is very rapid and clean and requires only a simple jig to hold the part in the correct position.

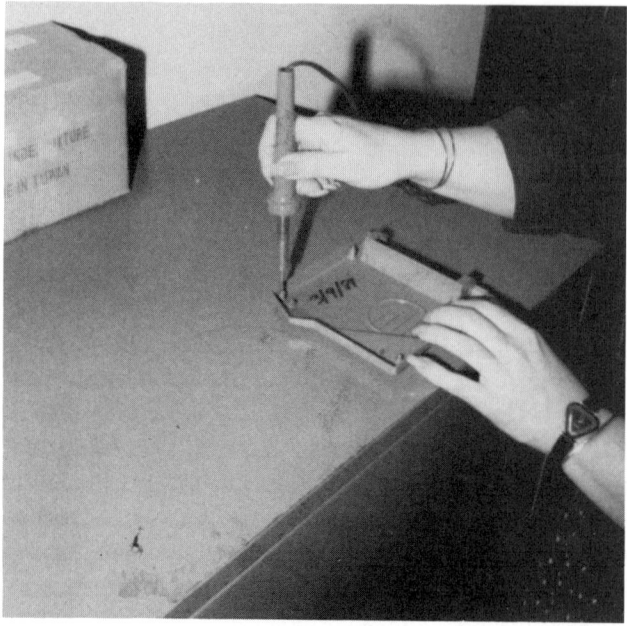

Figure 30
Installation of metal inserts by use of a hand-held soldering iron. This is quick and does not require extensive jigging. (Photo courtesy of Technical Molded Plastic.)

Adhesives. It is important to be aware of the new adhesive systems and their relationship to many plastic polymers. Using an adhesive is often the only method of joining two plastic parts or joining a metal part to a plastic part. Table 2 gives a list of some adhesives and the method by which they are cured.

MACHINING

Machining of molded parts is often necessary due to a design oversight or similar problem. Often, a file or router can be utilized, and a simple solution can sometimes be found in the use of a piece of sandpaper. Once in a while it is necessary to have parts sent to a machinist, where

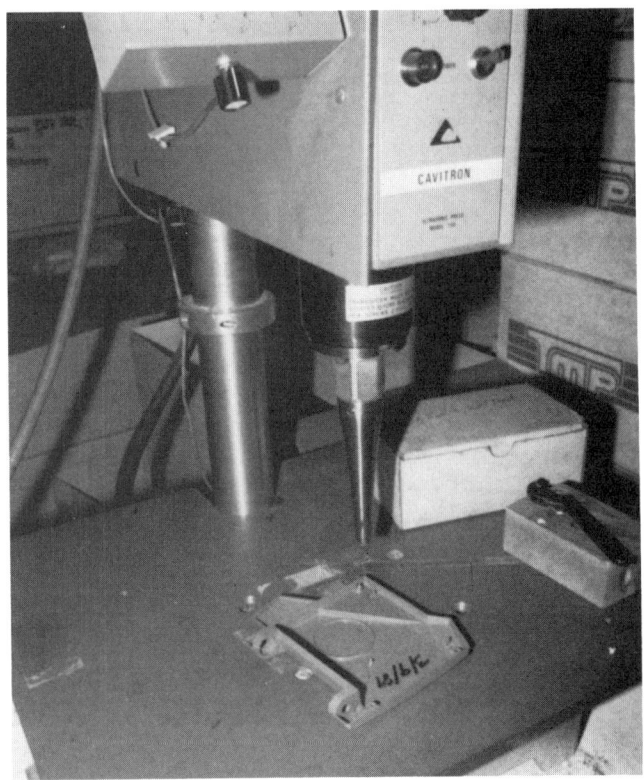

Figure 31
The use of an ultrasonic machine to install metal inserts into a molded plastic item. The proper jigging of the part assures the straightness of the insertion. (Photo courtesy of Technical Molded Plastics.)

they can be worked on a mill or lathe. This is obviously a very expensive step and should be avoided it at all possible.

The need to have holes drilled in a plastic part is often the subject of much confusion. If the hole can be cored during the molding operation, this is by far the least expensive method. Often this is not possible, such as where the hole is in a different direction from the tool opening. One way to accomplish this would be to have a cam-operated core pull or a hydraulic cylinder designed into the mold. This may be more

Table 1 Insert and Hole Dimensions[a]

Size of insert	Insert length	Hole size	Lead in hole
No. 0 and No. 2	0.115	0.121	0.126
	0.188	0.110	0.126
No. 4	0.135	0.156	0.162
	0.219	0.144	0.162
No. 6	0.150	0.202	0.209
	0.250	0.188	0.209
No. 8	0.185	0.229	0.237
	0.312	0.211	0.237
No. 10	0.225	0.270	0.280
	0.375	0.249	0.280
¼ in.	0.300	0.352	0.366
	0.500	0.324	0.366
5/16 in.	0.335	0.434	0.451
	0.562	0.404	0.451
3/8 in.	0.375	0.526	0.543
	0.625	0.491	0.543

Source: *Dodge Ultrasonic Inserts*, Heli-Coil Products, Division of Mite Corp.
[a]Minimum depth of hole should equal insert lengths plus 0.30 in.

expensive than just having the hold drilled in a post-molding operation. A rule of thumb is that a core pull built into a mold will cost about
$1,000.

Another question that often comes up is whether or not the application will require metal inserts. If the two parts being joined by use of
mechanical fasteners are going to be taken apart more than three or
four times over the life time of the product, threaded metal inserts should
be used. If not, a cored or drilled hole should retain the screw. This
decision will save approximately $0.10 per insert.

Flash removal is often a problem with molded and formed parts, and
this gets much more severe as the tool gets older. This is the responsibility of the molder, but often, the customer may be faced with the
question of paying for fixing up the mold or paying more for the molded
parts. Hand trimming of parts can cause problems with the appearance
of the parts and may not be satisfactory. When the situation develops,
one should have a meeting with the molder and get a rundown on all
the possibilities and their costs.

Table 2 Adhesives and Their Characteristics

Type	Name	Source	Cure	Comments
Polyurethane	Pliogrip II	Ashland Chemical	Air	4500 psi tensile, retains 80% of strength after water immersion
Methacrylate-modified urethane		Dymax Corp.	Air and UV	1–10 s cure, tolerates surface contamination
Water-based urethane	Silaprene M6184	Uniroyal	Air	Slow evaporation, used for vacuum forming
Urethane	Morthane PE	Morton Thiokol	Air	Flexible adhesive for bonding leather, vinyl, and elastomers
Epoxy	Fusor	Lord Corp.	Chemical	Two-part, less shrinkage
		Ciba-Geigy	Chemical	1-min cure, resists high temperatures

(continued)

Table 2 *Continued*

Type	Name	Source	Cure	Comments
	Moisture-Lok 101	Epoxy Tech.	Chemical	Absorbs little moisture, no solvents
	Scotchweld EC-348	3M	Heat	One-part adhesive, noted for its hot wet strength
	Cybond FT 5000	American Cyanamid	Heat	Twice the pot life (30 min), cures at 250°F
	EA 9461	Dexter Corp.	Heat	Bonds to plastics without a primer
	EA 9394	Dexter Corp.	Air	Room-temperature cure, bonds to oily substrates
	Araldit AV 8531	Ciba-Geigy	Air	Noted for flexibility
	Epaweld 13230	Hardman	Air	Cures at room temperature

Nylon–Epoxy		Tra-Con	Air	Tough flexible bond, solvent-free
Fluoroepoxy		Allied Corp.	Air	Needs no surface preparation, used for bonding fluoroplastics
Acrylics		Permabond Inter.	Air	Room-temperature cure, no solvents
	Versilok HI 400	Lord Corp.	Air	High peel and impact
Methacrylate		ITW Adhesive Systems	Air	High fracture toughness, bonds SMC to nylon
Acrylic		Loctite	UV or visible light	Requires no mixing no solvents
		Valchem	Radiation	Electron beam used for cure

(continued)

Table 2 *Continued*

Type	Name	Source	Cure	Comments
Epoxy–acrylic		Lord Corp.	Heat	High-temperature resistance, bonds SMC to metals
Cyanoacrylic	Chemlock	Lord Corp.	Air	Bonds rubber
	Aravite C13	Ciba-Geigy	Air	Get at room temperature, bonds in 15 s
	Superbonder 414	Loctite	Air	Improved bonding to ?
	Powerbond 920	Permabond	Heat	Can be used on polyolefins, withstands 475°F, fill-4-mil gap in 15 s
Hot melt (styrenic block)	Kraton	Shell Chemical	Cools	Fast setup, can be made to fail at a specific temperature

Type	Product	Manufacturer	Cure	Properties
Hot melt (urethane and epoxy)		W. R. Grace	Heat	Considerable impact
Hot melt (urethane)	Supergrip 2000	Bostick	Moisture	Resists creep, must be kept dry
Silicone		Loctite Corp.	UV	Withstands high temperature and weathering
	SEA 223	GE Silicones	Air	Does not require a primer, resists ozone and chemicals
		Norton Performance Plastics	Air	Dual-adhesive mounting tape
Polyiide–siloxane		Virginia Polytechnic Institute		High-temperature adhesive
Semicrystalline polyimide		NASA		High-temperature adhesive, unaffected by submersion in boiling water

Source: Allen J. King, "Update on Adhesives," *Plastic Design Forum*, May–June 1989, pp. 59–65.

Machining of Prototypes.   To build prototypes and models for various plastic projects, it is sometimes necessary to be familiar with machine techniques used in fabricating shapes from plastics. Basic shapes can easily be machined on all types of metal- and woodworking machinery. However, the low thermal conductivity, low melting point, and elastic behavior of plastics must be taken into consideration.

Carbon steel, high-speed tool steel, and carbide-tipped tools can all be used. However, the last two are preferred for long production runs. Tools must be kept sharp and smooth at all times and have sufficient side clearance angles. Use relatively high cutting speed and low feed rates. Coolants such as mist spray, water-soluble oils, and air be applied to machining operations only when excessive heat is generated and to improve chip removal.

Proper clamping and support for the worktable are required to avoid vibrations and consequent rough cutting. Sleeve bearings, especially those with a thin wall, should be supported on the inside diameter with a plug when clamped on the outside diameter by a multiple-jaw chuck. This will prevent distortion of the part during machining. All corners should be rounded to avoid failure due to the notch effect.

Machine Tolerances.   The machine tolerances for plastics are higher than those normally applied to metals. This is because of the increased coefficient of thermal expansion and possible deformations caused by internal stress relieving during and after machining. The latter phenomenon occurs primarily on parts where machining causes heavy section changes (e.g., on a bearing machine out of a large round rod). In these cases, thermal treatment (stress relieving) after machining of the part is necessary (minimum oversize, 3%) to obtain optimum dimensional stability and close tolerances. To date there are no international standards for machining tolerances of thermoplastic parts. In practice a machining tolerance of 0.1 to 0.2% of the nominal size can be allowed without taking special precautions (minimum tolerances for small dimensions, 0.002 in.).

Milling.   Milling cutters for light metals can be used. However, fly cutters are preferred because of the much better swarf removal.

Drilling. High-speed steel twist drills are recommended. Drilling is often the most critical operation as far as heat generation is concerned, and the use of cooling liquids is strongly recommended. To assist the cooling process and improve swarf removal, infrequent pull outs ("peck" drilling) are necessary, especially for deep holes. When drilling large holes, it is advisable to use drills with a thinned web to reduce friction and consequently heat generation. For large holes, step drilling is recommended; for example, a 2-in.-diameter hole should be made by drilling successively with 5/8-, 1 3/8-, and 2-in. drills.

Reaming. Properly sharpened straight-fluted or spiral-fluted reamers can be used. At least 0.004 to 0.101 in. should be left in the hole for reaming, providing a "bite" for cutting edges of the reamer, otherwise, due to the resilience of the plastic materials, there will be a tendency to compress under the pressure of the reamer, resulting in inaccurate cutting.

Sawing. To assure good chip removal, the use of bandsaws, circular saws, or reciprocating saws that have widely spaced teeth is recommended. Due to the tendency of plastics to "close in" behind the cutting edge, the saw blades should also have enough set to avoid excessive heat buildup and binding of the saw.

7

Tooling for Plastics

In this chapter we cover the important subject of tooling. Because this is usually the most expensive part of any project, we cover the subject in great detail. The subject of tooling varies with the molding or forming process you use, but the process used to obtain the individual mold is similar.*

TOOLING THE PLASTIC PROJECT

One of the most important and least understood parts of a plastic project is the tooling. This is true whether you are producing the part by injection molding or by some other process. First, it is probably considered the most expensive part of any project. Second, it is probably the least understood and the most complex. In fact, you are buying a complex piece of machinery when you purchase a plastic mold or tool, and normally you must rely on the expertise of the molder and toolmaker.

---

*Some of the following information has been taken from Borg-Warner Chemicals' *Design Tips, Number 18.*

Figure 32
A toolmaker works on the cavity of a large mold. (Photo courtesy of SPI.)

Cooperation Needed.   Close cooperation between the customer, toolmaker, and molder should be an obvious requirement of any project. However, this is not always the case, especially if a tight bidding procedure was in effect when the molder was selected. Often, the project is not yet fully defined and the customer is asking for a firm tooling quote on both the parts and the tool. This is nearly an impossible task for the molder. It is surprising how many molders will accept this situation and after a short discussion with the toolmaker, quote on the tool without really knowing what the customer requires.

New Purchasing Concepts.   Many companies bypass the normal purchasing procedure of competitive bids and select vendors by past experience and reputation. This allows them to sit down with the molder and toolmaker selected and design the tool with all the input available. This close cooperation assures that the customer will get a high-quality tool designed to do the job intended.

As commented on by Ed Galli, senior editor for *Plastics Design Forum*, in a recent article: "When a product design is etched in stone before requests for bids on the tooling are sent out, the result can be a complex mold or many tooling changes or both. Xerox, a well-known example, have renovated their procurement procedures reducing or eliminating competitive bidding and dramatically reducing their supplier base. Such a practice demands close early involvement with a qualified supplier" [1].

TOOL LIFE

Tool life is another subject that needs discussing. Just how long will a tool stand up to the high pressures and wide temperature swings experienced in a molding operation? Obviously, this depends on a good many things. Design of the mold, materials of construction, part design, and what materials are run in the mold all have an effect on the life of the tool. Just what constitutes "life of a tool" and how to specify this when buying a mold are important factors. A person responsible for purchasing tooling should be aware of these factors and should buy accordingly. Knowing what you are buying can mean the difference in whether a project turns out well or whether you have an expensive fiasco on your hands.

MATERIALS AND CONSTRUCTION

There are several materials of construction from which tooling can be built. They are, in decreasing order of resistance to wear and tear:

Hardened steel	S-7
Soft unhardened steel	P-20
Aluminum	7075
Epoxy	
Wood	
Plaster of paris	

Usually, the decision is made on the basis of the pressures involved in the process and how many parts are expected to be produced from the tool. For example, if one were going to have parts produced in structural foam, where the highest pressure to be found in the cavity is from 500 to 1000 psi, aluminum would be the choice. On the other hand,

Figure 33
Producing a mold can take hours of precise machining. Here a moldmaker puts the final touches to a large core for a large aluminum structural foam mold. (Photo courtesy of SPI.)

if you were to build a tool for a large-volume injection-molded application, you would certainly pick hardened steel. This would be not only because of the high volume wanted but because of the high pressures involved (8,000 to 10,000 psi).

MOLD DESIGN FOR HIGH-HEAT MATERIALS

The use of high-heat polymers such as PEEK and polyether imide has caused some concern about the proper design of molds for these materials. With mold temperatures at 300 °F, the entire system is running at considerably higher temperatures than what you would find with the more conventional polymers. For example, use of aluminum tools is questionable, even for prototype molds. This is simply due to the high heat of the materials being injected into them. Cases have been known where the leader pins and bushings have actually fallen out of a mold due to the differences in expansion values caused by the high heats.

Hot runner systems are being used with the high-heat polymers, but careful attention must be paid to the degree of control available with the system, and careful consultation should be carried out with the material supplier, mold designer, and hot runner system supplier.

GUIDELINES FOR ORDERING TOOLING

A guideline to injection molds follows.*

The following classifications are guidelines to be used in obtaining quotations and placing orders for a uniform type of tooling. It is our desire through these classifications to help even out inequities in the plastic tooling quote system and eliminate customer disappointment.

It is strongly recommended that tooling drawings be obtained before construction is started on any injection mold. Even though parts may seem simple enough not to warrant tool design, in the event of tooling damage, a drawing showing sizes and steel types will pay for itself. These specifications refer to molds for standard thermoplastic materials and product generally not exceeding 30 in.² of projected surface area. Those parts would normally be run in machines with clamp ratings of 400 tons and less. Larger molds quite often have different requirements.

These classifications are for tooling specifications only and in no way guarantee workmanship. It is very important that purchasing agents deal with vendors whose workmanship standards and reliability are well proven.

Because of variations in materials, design, and molding conditions, mold life cannot be guaranteed. We attempt to give approximate cycles for each type of mold, excluding wear caused by material abrasion.

Tool Maintenance. Maintenance is not the responsibility of the moldmaker. Normally, maintenance such as replacement of broken

---

*The material on pp. 121–128 is from the booklet *Classification of Injection Molds*, C. Brewer Co., Anaheim, CA.

springs, broken ejector pins, worn rings, or the rework of nicks and scratches should be borne by the molder (provided that they originally built and have maintained possession of the tool). Maintenance caused by excessive wear should be borne by the mold buyer. Examples of this would be worn gates, worn cavities, and cores caused by gate "jetting" with abrasive materials; a flashing down and around ejector pins or sleeves caused by an excessive number of cycles; replating, or retexturing of cores; and/or cavities worn by abrasive materials. The possibility of mold rework costs should be considered carefully when deciding which classification of tooling is required.

GUIDE FOR PURCHASER

Classification of Injection Molds.   This section contains a brief synopsis of the various tooling classifications. The following section contains detailed descriptions of each tooling class. It is our recommendation that a tooling information sheet be included with each tooling request for quotation.

Class I Mold

Cycles: 1 million or more
Description: Built for extremely high production. This is the highest-priced tooling and is made with only the highest-quality materials.

Class II Mold

Cycles: Not exceeding 1 million
Description: A medium- to high-production mold, good for abrasive materials and/or parts requiring close tolerances. This is a high-quality, fairly priced tool.

Class III Mold

Cycles: Under 50,000
Description: Medium-production mold, a very popular mold for low- to medium-production needs. Most are in the common price range.

Class IV Mold

Cycles: Under 10,000

Description; Low-production mold. Used only for limited production, preferably with nonabrasive materials. Low to moderate price range.

Class V Mold

Cycles: Not exceeding 500
Description: Prototype only. This tooling will be constructed in the least expensive manner possible to produce a very limited quantity of prototypes.

Inserts. When buying mold inserts, the customer buys ony the insert. The unit mold base is owned by the molder. Because of the large variation in insert sizes, it should be kept in mind that it may be impossible to have the product produced by another vendor without having to purchase a mold base:

Class I Unit Insert

Cycles: Approximately 250,000
Description: Top-quality materials for medium- to high- production requirements.

Class II Unit Insert

Cycles: Under 50,000
Description: Similar to class III molds. Most commonly used insert. Low- to medium- production requirements.

Class III Unit Insert

Cycles: Less than 500
Description: Similar to class V mold. Least expensive for very limited quantities. Insert built with the least expensive materials.

GUIDE FOR MANUFACTURER

Classification of Injection Tools.   In this section we give details of materials and the processes to be used in producing the various classifications of tooling.

General Specifications

1. Customer approves mold design for class I, II, and III molds and class I and II inserts prior to the start of construction.
2. All tools, with the exception of the prototype, must have adequate channels for temperature control.
3. Wherever feasible, all details should be marked approximately by 0.005 in. deep with steel type and Rockwell hardness.
4. Customer name, part number, and tool number should be steel-stamped on all molds and/or inserts.
5. All molds should have eyebolt holes on the top side. There should also be one bolt hole above and one below in parting line to facilitate mold removal, if required, in halves.
6. The ejector plate ends should be painted high-visibility orange to comply with U.S. Occupational Safety and Health Administration (OSHA) standards.
7. All ejector pins should be nitrided or hardened.
8. Tie straps should be provided with molds to eliminate the possibility of damage caused by accidental opening of the tool.
9. Molds for materials that give off caustic or acidic fumes should have molding surfaces plated or made of steel resistant to those gases.
10. In multicavity molds, all identical cavities must be individually indentified, if possible.
11. Design of some molds requires that pins, blades, sleeves, or stripper rings needed for ejection be placed in front of the side covers. These molds must be built to allow hydraulic ejection return, or a return system should be installed for mold protection.

Class I Mold

1. Detailed tool design required. Should be approved by the customer and molder prior to the start of construction.
2. Mold base to be a uniform hardness of 300 Brinnel hardness number (BHN).
3. Molding surface (cavities and cores) must be tool-steel hardened to at least 48 Rockwell C. All other details, such as slides, heel blocks, gibs, and wedge blocks, should be hardened tool steel.
4. Ejection must be guided.
5. Slides must have wear plates.

6. Temperature-control channels should be in cavities, cores, and side cores whenever possible.
7. Cavities should be plated for protection unless stainless steel is used.
8. Hardened runner plates are required except for hot or insulated runner molds.
9. Electroless nickel plating of all water channels is recommended. This greatly inhibits the chance of rust and makes it easy to clear sediment from plugged lines.
10. Eyebolts are required on all four sides above and below the parting line.
11. Taper locks or similar interlocking devices are required on molds where extremely close tolerances must be maintained.
12. It is recommended that the water-bearing plates be electroless nickel-plated, parkerized, or otherwise protected from rust or corrosion.

Class II Molds

1. Detailed tool design required. Should be approved by the customer and molder prior to the start of construction.
2. Cavity, core, stripper, retainer, and backing plates to be a uniform hardness of 300 BHN.
3. Molding surfaces (cavities and cores) must be tool-steel hardened to at least 48 Rockwell C. All other functional details should be produced and heat treated.
4. Temperature control channels should be directly in the cavities, cores, and side cores wherever possible.
5. Taper lock or similar interlocking devices are required on molds where extremely close tolerances must be maintained.
6. The following items may or may not be required, depending on the ultimate production quantities anticipated. It is recommended that those items desired be checked and made a firm requirement for quoting purposes: (a) guided ejection, (b) slide wear plates, (c) plated temperature control channels, (d) plated cavities, and (e) hardened runner bars.

Class III Mold

1. Detailed tool design recommended.

2. Cavity backup plates must have a uniform hardness of 165 BHN.
3. Cavity and cores must be 300 BHN or higher.
4. All other extras are optional.

Class IV Mold

1. Tool design recommended.
2. Mold base can be of mild steel or aluminum.
3. Cavities can be of aluminum, mild steel, or any other agree-upon metal.
4. All other extras are optional.

Class V Mold

1. May be constructed from cast metal or epoxy or any other material offering sufficient strength to produce minimum prototype pieces.

Inserts

Class I Unit Insert

1. Detailed tool design required. Should be approved by the customer and molder prior to the start of construction.
2. Insert to be a uniform hardness of at least 300 BHN.
3. All molding and/or functional details are to be made of tool steel hardened to at least 48 Rockwell C.
4. Slides must have wear plates.
5. Temperature control channels to be in cavities and cores wherever possible.
6. Molding details should be plated for protection unless stainless steel is used.
7. Electroless nickel plating on all water channels is recommended.
8. Taper locks of similar interlocking devices are required on inserts where extremely tight tolerances must be maintained.
9. Inserts must have leader pins and bushings or a similar guidance system.
10. It is recommended that the entire insert be electroless nickel-plated, parkerized, or otherwise protected from rust or corrosion.

Class II Unit Insert

1. Detailed tool design required. Should be approved by the customer and molder.

Figure 34
These photos show two types of mold inserts that are used with a mold chase
or base. The molder puts these inserts into the common frame, which provides
a foundation for the inserts and allows their installation into a molding machine.
Many sizes are available, and this becomes an efficient way to buy tooling,
especially for small parts. (Photos courtesy of Technical Molded Plastics.)

2. Insert retainer to have a uniform hardness of at least 165 BHN.
3. Cavities and cores must be 300 BHN or higher.
4. All other extras optional.

Class III Unit Insert

1. Can be constructed from aluminum, cast metal, cast epoxy, or any material with sufficient strength to produce at least 55 injection-molded pieces.

ECONOMICS

Economics often plays an important part in selecting the material to be used for production of the tooling. This should not be allowed to happen. Material of construction should be selected entirely on the basis of what material is best for a given process situation. You should not pick a material that is too good or too expensive for an application. And you certainly do not want a material that will not hold up to a given set of conditions. If you have selected the right material for the job, the worst thing that can happen is that your tool will need some minor repair work after it has produced several thousand parts.

PLASTIC TOOLING

Much has been said about plastic tooling, and a great deal of work has been done with it in recent years. Such companies as 3M offer a service to provide cavities and cores for an injection mold. In a recent book William Benjamin notes: "The use of plastics as tooling materials has provided many new opportunities for the creative plastic engineer and tool engineer to advance the state of the art and improve tooling costs...Through the continued development of new tool fabrication techniques the plastic tooling industry has steadily grown. Tooling is a prime consideration in the decision making processes involved in the planning of the manufacture of any product. It is obvious that because the tooling costs are part of the manufacturer's investment, these costs must be recovered as part of the price of the product" [2]. Therefore, it is very important that the lowest-cost material that will do the job be chosen.

One could choose plastic tooling and be well satisfied. However, one could also get a plastic tool that would fail on the first cycle and lose money and, more important, precious time. Plastic tooling has been used in some types of plastic processing. One should investigate all the ramifications before choosing a material of construction on the basis of price alone. The best sources of information are the molder and toolmaker.

TOOLING DESIGN

Tool design and who is responsible for it is another important subject. Many large companies have their own tooling design engineers. Molders and toolmakers like nothing better than to be handed a complete tooling package, complete with a worked-out tool design, as it removes all responsibility from them. However, most of us are not lucky enough to have this type of service available. We must get the tooling design from somewhere else. Usually, we rely on the chosen molder and toolmaker.

DEALING WITH MOLDMAKERS

The need for patience when dealing with a moldmaker or molder is important to the outcome of any project. During the duration of a project with a business machine manufacturer the toolmaker had run into problems, and the tool, a housing, was late. The project engineer on the job was informed and immediately became very irate over the slipped schedule. Not much could be done about improving the delivery date, but you could not tell the engineer this. He fumed and fussed for two weeks and finally threatened to cancel the mold. His anger no doubt could have given him an ulcer if he was so inclined.

The mold finally came in and the first article samples were submitted. However, the engineer had become so upset by the delay that to this day he refuses to speak to either the molder or the moldmaker. Schedules are important, but sometimes things happen to influence deliveries, and if you are wise, you will accept them and make allowances for the delay.

TOLERANCES

Tolerances can be very important in the development of a tool to be used for the production of a plastic part. When a truck manufacturer

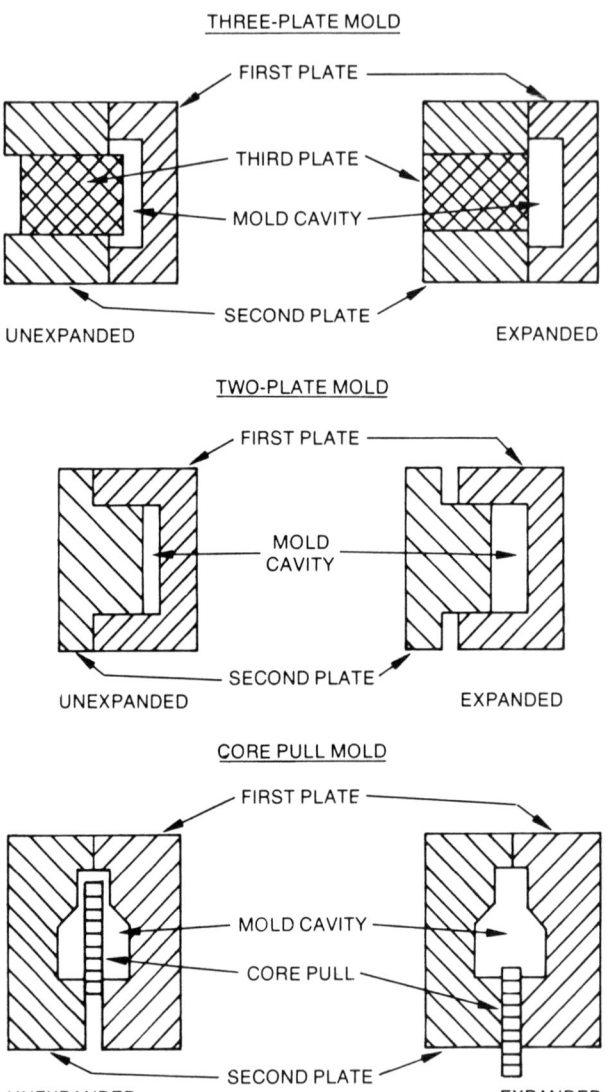

Figure 35

Three types of simple mold construction. Various configurations of these three systems are utilized in moldmaking.

recently had a large structural foam tool made for the production of a full dashboard, the philosophy of tolerances was not discussed. The toolmaker produced the tooling to one set of tolerances but they were not what the original equipment manufacturer (OEM) had in mind. The tool came in, samples were molded, and dimensions were measured. The truck manufacturer was insistent on having every dimension on the print right to tolerance. This amounted to 5000 to 8000 dimensions. The toolmaker admitted having some problems with the dimensions but wanted relief on the noncritical ones. The OEM stuck to their original demand and an impasse soon developed.

To meet the manufacturer's requirement, the tool would have had to be made over completely. Finally, cooler heads prevailed and the part was installed in a truck. From this it was learned which dimensions actually had to be changed to allow the part to fit properly. The toolmaker corrected these dimensions and the part was declared "fit for production." The engineering department later changed the drawing dimensions to match the part. Closer coodination between toolmaker and customer would have prevented this expensive loss of time. The important thing to learn here is not to put excessively tight tolerances on your drawing unless they are actually needed. Tight tolerances cause the cost of the tool to go up, and often the molder cannot hold the tolerance anyway. It is a good idea to have a discussion about this subject before the contract for the molding or the tool is let. Any agreement reached should be made in writing, and all parties should know exactly what they are agreeing to do.

REFERENCES

1. Ed Galli, "How to be Kind to Your Toolmaker". *Plastics Design Forum*, Sept.–Oct. 1987.
2. William Benjamin, *Plastic Tooling Techniques and Applications*, McGraw-Hill Book Co., New York 1982, pp. 1–2.

8

Dealing with the Plastics Industry

DEALING WITH THE PLASTICS INDUSTRY

Like any industry, the plastics industry has its own peculiarities. Many types of people work on the fringes of the industry and help to assist potential users of plastic materials. Everyone from product designers to patent lawyers are important in their own way. Many of the suppliers who serve the plastics industry also have their own way of operating, and in this chapter we assist you in dealing with them. The custom molder, for example, fills an important niche in the plastics industry. Many of them are small but provide a very necessary service by molding a wide variety of plastic applications for a large cross section of industry.

Other items are covered in this chapter to assist engineers in dealing with the plastics industry. Whether to write a marketing plan and the pricing of a product are other subjects covered.

USING A PRODUCT DESIGN

Often there is a question of whether to use a product designer. You have a basic idea of what you want to build; perhaps you even have a working model. Do you really need to spend money for a designer?

First, you are going to need engineering drawings of your product. These are necessary because the moldmaker must have them before building the mold, and the molder must have them to quote a price and know just exactly what is being made. You need them so that there is a thorough understanding of just what you are buying. Without them, you have only conversation; with them you can always refer back to them and know exactly what it is you are buying.

Unless you are an expert drafter, it is well worth the money to get an experienced person to draw up the prints. Sometimes it is economical to enlist the services of a mechanical drafter rather than a product designer. This will work only if you have a good idea of your product and know just what it is you are going to build. Using a product designer often helps to improve the product by getting someone with fresh ideas involved in the development process. Sometimes the help of someone with a knowledge of mechanical engineering is needed. This should be considered when hiring a design engineer. Make sure that the designer has the qualifications you need before starting the design process.

The important thing is to have available to you engineering drawings that can be read and understood by all concerned. These drawings should be gone over by somebody who understands plastic design. If the person who drew up the drawings is knowledgeable about plastic design, so much the better. Otherwise, let your molder look them over and comment on them. It is often wise to have a material supplier look them over also. Suppliers often provide this service free of charge and can give you valuable hints on the project. The telephone numbers of several material suppliers who can assist in design and material selection are given in Appendix III.

Questions That Should Be Asked by a Designer.   All designers worth their salt will have certain questions to ask you before starting a project. These include the following:

1.   What does the part do?
2.   How many component parts do you think will be needed in the assembly?
3.   Under what environmental conditions will the part operate?
4.   What are the highest and lowest temperatures in which the part will have to function?

5. What volumes do you plan to produce the first year, the second year, in later years?
6. What process do you plan to use to produce your product?
7. What material do you plan to use in your product?
8. What inserts will be used in your product, and how are you going to install them?
9. Is there a need for post-molding operations, such as drilling or machining?

These and other questions will need answers before the designer can do an effective job of designing the product.

TOLERANCES

Together with engineering drawings should be a selection of tolerances needed to produce the product correctly and with the least amount of difficulty. These will vary some according to the process you select to manufacture the product. This is one of the things you should discuss with the molder. However, once you decide on a working tolerance, make sure that it is incorporated into the drawings. It may be one of the things that you will need as back up if you have problems with a nonfunctioning product.

Tolerances can be very important in the development of a plastic product. The material shrinkage has to be taken into consideration when choosing tolerances of molded plastic parts. These shrinkages vary according to the material chosen. This will also vary if the material callout includes such additives as glass or carbon fibers. There is a tendency on the part of most design engineers to put too tight a tolerance on their part designs. This can be as harmful as not using a tight enough tolerance or no tolerance at all. Tolerances may be so tight that they are not within the capability of the process being considered. The toolmaker may have no problem holding $\pm$ 0.001 in. on the molds, but if the molder is unable to hold this dimension, it is unlikely that so tight a tolerance will be met.

A truck manufacturer recently had a large structural foam tool made by a molder who was going to produce a fairly complex dashboard for the manufacturer. The pros and cons of specifying close tolerances were not discussed before the project got under way. The toolmaker produced the tooling to one set of tolerances, but they were not what the

OEM had in mind. The molder could not hold the tolerances that the manufacturer demanded. The first article was produced and dimensions checked.

The OEM was insistent on the need for every dimension on the print to be within tolerance. This amounts to some 8000 dimensions—some critical, others not as critical. The moldmaker admitted to having problems with some of the dimensions but wanted some relief on the noncritical measurements. The OEM stuck to their original demand and an impasse developed with the molder in the middle. To meet the truck manufacturer's requirement, the tool would have to made completely over again, and the moldmaker was not willing to do this. Finally, cooler heads prevailed and sample parts were installed on the truck. From this it was learned which dimensions actually had to be changed to make this part fit on the vehicle. The toolmaker then corrected those dimensions and the parts were declared "production." The engineering department later changed the drawings to match with what was actually being produced. Closer coordination between toolmaker and customer would have prevented this "discussion" and with it the loss of valuable time.

A tolerance callout often used in the plastics industry is as follows:

Two-place dimensions (0.XX) $\pm$ 0.01 in.
Three-place dimensions (.XXX) $\pm$ .005 in.

On larger dimensions the tolerances are usually relaxed, while on smaller dimensions they can be tightened up. You should talk this subject over with the molder before defining final tolerances. To put tighter tolerances on the dimensions than you need is asking for problems. First, it costs additional money for the molder to try to hold tighter tolerances than necessary. These costs get passed on to you, the customer. Also, some processes just are not capable of tight tolerances and you will get parts that do not meet the specifications anyway.

WORKING WITH A VENDOR

Many of the problems encountered in a plastic project come from a failure to communicate properly. This situation usually starts when the customer meets with the vendor for the first time, and it often continues through the life of the project.

Be sure that you have an understanding of what you want, then be sure you tell the vendor exactly what you want and put it in writing. From the engineering drawings to the order for tooling and parts there is much room for misunderstanding. Putting it in writing protects you as well as keeping mistakes to a minimum. Plastics projects are like any other project where a million details are necessary—it takes only a few wrong ones to spoil the project.

Some things to put in writing include:

1. Details of the part you want made.
2. Material from which you want the part made.
3. Color that you want the part to be.
4. Number of parts you want produced.
5. Details of the tooling you expect to buy.
6. Who is to furnish any hardware that may be needed.
7. A complete schedule of delivery on both the tooling and first articles.
8. Any packaging requirements, and who is to provide them.
9. What markings are to be put on the box.
10. How the products are supposed to be shipped, and to whom.

This and other information will be necessary to get the project started and to ensure that you get exactly what you want.

TIMING ON A PROJECT

Timing can defeat a project before it starts. Even though the project engineer can start out with the best intentions, the project can be doomed if the engineer is misled about timing or choses to ignore timing.

One should first establish reasonable goals. These would be such things as prototypes available, tooling complete, or first production available. These should be established by a careful study of what is to be accomplished.

Once you have done this, it is time to set up reasonable scheduling to get the project done within the time frame you have established. One can hope for a miracle, but it seldom happens. Give your suppliers enough time to accomplish what they need to do, and then add some slack for unforeseen problems.

Take the building of the tool, for example. The molder will give you a delivery schedule based on the time established by the toolmaker.

Setting up a weekly report system is a good way to make sure that the molder is kept aware of how the tool is coming along. Another thing that helps assure you of on-time delivery is a bonus for days knocked off the original delivery date, and if you can get the molder to go along with it, a penalty for each day the toolmaker is late.

You should figure on at least 2 weeks' time after mold delivery for "tweeking" the mold: that is, getting the mold into shape for a production run. You can count on it taking at least this long to make changes you want as well as changes requested by the molder. Some tools will take less time, others more. There is no benefit in pushing a moldmaker beyond their usual pace of business. Undue pressure only succeeds in pushing the moldmaker into making bads mistakes that cause you to lose more time. Remember that the toolmaker wants to get the mold done as badly as you do and does not get paid until the tool has produced acceptable production parts.

Timing is also important when you are trying to get a product ready for a specific period of the year. Often in trying to get a product ready for a particular season, you put undue pressure on the moldmaker, molder, and others. Allow enough time for everybody involved to get their parts of the development finished. You may miss one season but you will eliminate problems that could cost you more time and money if the project fails.

DESIGN AND MATERIAL SELECTION ASSISTANCE

The best sources of information on design and material selection are often the various material suppliers. In many cases they can point out areas that need changing simply by looking at the drawings. A list of suppliers' telephone numbers is given in Appendix III. A call will probably bring you a wealth of good technical assistance, usually at no cost.

According to an editorial by Mel Friedman in *Plastics Design Forum*: "Most designers say that the biggest drawback to their using plastics, even more than they already do, is a lack of sufficient design information or knowledge about these materials, or understanding of how they work. Many of the basic resin producers and compounders are willing today to go to great lengths—well beyond property data sheets and design manuals—to help in design work: many offer state of the art design support services, [1]. Some of the companies offering such help are Dow, Monsanto, LNP, Borg-Warner, USI, and GE.

Part Design

The second step in the development of any project is the design of the product. It is not enough just to think up a better way to do something: One has to have the item developed to the point where it can actually be produced. This means design and all that it entails. Borg-Warner Chemicals defines design as "a mental project or scheme in which means to an end are laid down." [2]: a preliminary sketch or outline showing the main features of something to be executed; the arrangements of elements that make up a work of art, a machine, or other manufactured project: deliberate purposive planning. Within the plastics industry, the term *design* means different things to people in different areas of specialization.

To the toolmaker, *design* means the layout and construction of tooling to incorporate production efficiencies, good tool durability, and economies of cost.

To the processor, *design* means planning the availability of equipment for production of a product with optimum efficiency and economics of cost.

To the project engineer or product designer, *design* means integration of end-use performance requirements and material property limits into development of the part design. For all these people three factors appear fundamental to the process of design, whether designing a tool, a process, or a product:

1. Basic understanding of the objective
2. Good planning
3. Adequate background information

To design a new product, the designer must understand the basic objective—the function of the product. The designer must schedule an adequate portion of the product development timetable for in-depth background information on the end-use performance requirements of the application.

Plastics are typically used closer to their design limits than are metals or wood. The performance of a product reflects the designer's understanding of the behavioral characteristics of the plastic, as well as the mechanical properties of the plastic grade used. Ultimately, the challenge to the designer is to ensure the material behavior, and performance demands are complementary in the final product. Careful attention to detail is important because design has yet another meaning: To the end user, *design* mean reliable part performance through normal, every-day use.

The Borg-Warner manual [2] goes on with a five-step procedure for the development of a new product design:

Step 1 is the establishment of end-use requirements. Specific details needed to establish end-use requirements include: (a) anticipated structural requirements, (b) anticipated environments, (c) desired cost limits, and (d) regulation/standard compliance.

Step 2 is the development of a preliminary design. A preliminary design sketch will show the designer what aspects of the design are inflexible and which can be used to obtain the desired level of structural performance. The preliminary design sketch should therefore not be fixed with variable dimensions.

Step 3 is the selection of a plastic that will initially satisfy defined end-use requirements.

Step 4 is the modification of the design to reflect the specific property balance of the plastic material selected, the processing limitations, and finally, the assembly methods.

Step 5 is the testing of simulated use and simulated storage of the product. For product performance tests, specialized test equipment and techniques can be used, such as:

1. Strain-gauge analysis
2. Brittle coating analysis
3. Environmental chambers for the thermal cycling of parts
4. Infrared light banks for radiant-heat-effect determinations
5. Life testing under simulated use conditions
6. Accelerated aging under elevated temperature, high humidity, or UV exposure conditions
7. Creep testing using sustained loads

In the following paragraphs you will find that design and in-use tests should not be taken lightly. Many projects have fallen by the wayside because project managers did not take the necessary time to complete testing or to proof a product before it went to market.

DESIGNING FOR ASSEMBLY

Design is the first stage of manufacturing. Recent studies of computer-related products have shown that reductions of 20 to 40% in manufacturing cost and increases of 100 to 200% in assembly productivity are readily obtainable through proper consideration of assembly at the design stage [3]. This is an important point often forgotten by project developers in their haste to get a product off the drawing board and into the marketplace. The time to be aware of assembly techniques is during the design stage, not after the project has been tooled and you are stuck with poor and expensive assembly.

According to Boothroyd and Dewhurst [3], the reason that early assembly process selection is important is that manual assembly differs widely from automation assembly in the requirements it imposes on product design. What must be known before assembly requirements can be established is the projected market life, the number of parts to be produced, the projected production volume, and the company's investment policy.

The cost of assembling a product is related to both the design of the product and to the assembly process used for its production. Assembly cost is lowest when a product is designed so that it can be assembled economically by the most appropriate process. As outlined in the article by Boothroyd and Dewhurst, five basic types of assembly must be considered:

1. Manual assembly with mechanical assistance
2. Manual assembly on a multi-station assembly line
3. Automatic assembly using manually loaded part magazines and sophisticated two-armed robots with a special-purpose gripper that can handle all the parts for a single assembly
4. Automatic assembly using a manually loaded part magazine and a free transfer machine with programmable workheads, capable of performing several assembly tasks
5. Automatic assembly using special-purpose free transfer machines, workheads, and automatic feeders

The subject of automation as it applies to assembly is beyond the scope of this book. Readers who wish to obtain more information on this subject should obtain the *Design for Assembly Handbook*, developed by Boothroyd and Dewhurst.

DESIGN CONSIDERATIONS

A large amount of information is needed by designers before they can develop a product design from its initial concept. A project manager should be prepared to have this information available before contacting a designer. Some of this information may not be readily available, but an effort should be made to obtain as much information as possible since the absence of an item may cause a severe deficiency in the design. We describe the next five steps to acquiring recommended items of information that should be available to the designer.

Step 1 is to establish end-use requirements. In other words, what does the product do? Material selection must be keyed to the specific properties critical to part performance. The specific details needed to establish end-use requirements include:

1. Anticipated structural requirements
   a. Loads
   b. Impact forces
   c. Rate of loading
   d. Duration of loading
   e. Foreseeable misuse
   f. Vibration and fatigue
2. Anticipated environments
   a. Temperature extremes
   b. Exposure to chemicals, such as cleaners
   c. Exposure to sunlight and other UV sources
   d. Weather protection required
3. Desired cost constraints
   a. Cost of the product desired
   b. Anticipated yearly volume
   c. Processing method economically feasible
   d. Price of competitive materials of construction
   e. Expected service life and replacement interval
4. Regulations and standards compliance
   a. UL: Underwriters' Laboratories
   b. CSA: Canadian Standards Association

    c. NAHB: National Association of Homebuilders
    d. IAPMO: International Association of Plumbing
    e. FAA: Federal Aviation Administration
5. Mechanical officials
    a. FMVSS: Federal Motor Vehicle Safety Standards
    b. ANSI: American National Standards Institute
    c. ASTM: American Society of Testing and Materials
    d. PPI Plastics Pipe Institute
    e. FDA: Food and Drug Administration
    f. NSF: National Sanitation Foundation

Step 2 is to establish a preliminary design. A preliminary design sketch will show the designer what aspects of the design are inflexible and which can be used to obtain the desired level of structural performance. With the preliminary sketch the designer needs to know both fixed and variable dimensions.

Step 3 is the selection of a material that will initially satisfy the defined end-use requirements. The typical property data sheet is the initial document the designer should review to select a material based on the defined end-use requirements.

The design engineer must refine the initial selection by reviewing the time-dependent, temperature-dependent, and environment-dependent property behavior as appropriate to application needs. Supplementary data on abrasion resistance, tensile elongation at break, and so on, may be needed to confirm the material choice.

Material properties can be divided into two categories:

1. Properties useful for mathematical engineering design
    a. Proportional limit
    b. Modulus versus temperature
    c. Apparent (creep) modulus
    d. Fatigue endurance limit
    e. Poisson' ratio
    f. Stress rupture
    g. Coefficient of expansion
    h. Coefficient of friction
    i. Specific gravity
    j. Mold shrinkage

2. Properties that indicate part performance but require some ex-
    perience to relate the meaning of the property to application
    a. Hardness
    b. Impact strength (Izod or falling dart)
    c. Chemical resistance
    d. Tabor abrasion
    e. Weather resistance
    f. Tensile elongation at break
    g. Flammability characteristics
    h. Heat deflection temperatures
    i. Tensile strength at yield and/or fail
    j. Electronic properties

Step 4 is to modify the design to reflect:

1. Process limitations
2. Assembly methods

Prototyping. At this point in the design development effort, a prototype should be developed from the latest blueprints. Prototyping and
prototype testing can help the designer by:

1. Establishing confidence in the design by determining that re-
    quirements do not exceed design limits
2. Developing preliminary product performance information
3. Identifying potential problem areas

Four aspects are essential for successful testing. These are:

1. Good analysis of application end-use requirements
2. Close similarity to the final product, particularly in critical and
    suspect performance areas
3. Development of a realistic simulated-use or simulated-storage testing
    procedure
4. Commitment to the time and effort required for testing before prod-
    uct introduction

Step 5 is to conduct simulated-use product performance tests. It is important to determine which part performance tests reflect on defined
end-use requirements and can be conducted in the laboratory. Frequently,
the test equipment developed can be used for future production quality
control testing, which is an added benefit.

Product performance tests can be conducted on the functional prototypes or on production parts. Since functional prototypes can be produced using prototype, nonproduction tooling or part modeling, some caution must be exercised during testing. The prototype may not respond the same as will a production part. The initial production parts should also be tested, to confirm product performance tests.

CHECKLIST FOR A PLASTICS PROJECT*

In the development of any plastics project, one should use a checklist such as the following to assure that all items are taken care of.

1. Develop an idea.
2. Put the idea down on paper. This can be a simple sketch, but remember to date it and save it.
3. Make up a simple working model. It does not have to be constructed of the right material but must be good enough to prove that the idea will work.
4. Try out the model in an in-use test. Again, be sure to document the trial and retain all information.
5. Make any changes to your model and develop an engineering drawing of the part.
6. Discuss the project with a plastic processor and get an idea of various ways in which the part can be produced.
7. Get preliminary quotes on tooling and production cost for all feasible processes.
8. Choose a production technique.
9. Choose a material or materials that will meet your requirements.
10. Get a final quotation on tooling and production.
11. Choose a vendor and place an order.
12. Set up a delivery schedule and reporting system with the vendor.
13. Develop packaging and complete the purchase of all other items, such as inserts, fasteners, jig fixtures, and drill fixtures
14. Take delivery of tooling and molded first articles.
15. Receive the first article report from the vendor.
16. Have corrections or changes made in the mold.

---

*Checklist prepared by Fred Brieninger, a project engineer for Mannesmann Tally, a computer printer company located in Seattle, Washington.

17. Run a small volume of production parts.
18. Set up quality criteria on the parts.
19. Accept the first production run of parts.

This procedure will vary slightly depending on the process used; the complexity of the part, and any other factors.

WHAT TO EXPECT FROM A VENDOR

When you finally decide on a vendor, what can you expect from them, and what can they expect from you? First, of course, you can expect a finished plastic part, complete to your specifications. It should be made of the material you have specified and in the completed form described in the sales agreement.

Most molders agree to furnish a tool that will mold a part to prints or specifications. Once the tool is available, you should expect to receive production parts from that tool in volumes defined in the purchase order.

What can the molder expect from you? First, you should supply them with readible drawings complete with material callouts and any other requirements that you expect from the vendor. Tolerances should be spelled out. It should not be up to the molder to decide what material to use or what color the parts should be. The molder will often give you suggestions, but your must make the ultimate decisions.

Communication is the key word here. You should know what to expect from your molder, and they should know what to expect from you. Sometimes there has to be a little give and take. Most molding processes require a little luck. Sometimes a part will not come out exactly as you expect. That is when you need to work with the molder and see if you cannot, together, come up with a plastic part that will do the job.

EVALUATING A PROSPECTIVE PLASTICS VENDOR

Once you have established that process you want to use, it becomes necessary to evaluate several vendors, to enable you to choose the best one. Following is a checklist to use in evaluating a prospective vendor:

1. Plant location
    a. Location of the plant where work will be done.
    b . Trucking services to the plant.
    c . Rail and air service to the plant (both freight and passenger).

    d. Source of electrical power. Discuss outages, dependability, plant load, and so on.

    e. Water supply, water treatment, and general quality.

    f. Other things of interest, such as environment surrounding plant, industrial area, and unpaved roads.

All impressions, positive and negative, are worth noting, especially on the first visit.

2. Plant ownership and interests

    a. How is the company held: privately, partnership, stock, subsidiary, or other?

    b. How long has the organization been in the plastics molding business?

    c. What proprietary products are produced in the plant?

    d. What custom products are produced?

    e. What is their preferred work? Is there a greater expertise in one area?

3. Personnel

    a. Is it a union plant? Have there been problems? What kind?

    b. List and describe management.

    c. Define the professional and technical staff.

    d. Describe the chain of command and organizational chart.

    e. Define the quality control procedures.

    f. Define any health and safety problems.

    g. Describe the training programs.

    h. Give a general assessment of plant morale.

4. Plant characteristics

    a. Are the buildings protected against fire?

    b. What is the general layout? What is the material flow?

    c. Define the equipment arrangement. Is the area crowded? Is space allowed for accessory equipment?

    d. Is material storage planned, or improvised?

    e. Define how finished products are handled. How are they stored and moved?

       (1) Are facilities available for placement of inserts?

       (2) Are painting facilities available off-site?

    f. Shipping: How are parts handled in preparation for shipping?

    g. Maintenance facility: Is it a planned operation? Is preventive maintenance practiced?

    h. In general, look at:
       (1) Ventilation
       (2) Heating–cooling
       (3) Lighting
       (4) Housekeeping

5. Machines
    a. Molding machines
       (1) Make and model numbers
       (2) Size
       (3) Age
       (4) Controls: type (e.g., computerized)
    b. Cooling systems: How are they controlled? Are there any temperature limitations?
    c. Dryers: kinds of driers and types of controls; how are they tested, how often, and by whom?
    d. Tooling storage: Is space reserved for each customer's tooling? Is the storage area in a sheltered building? What are the general conditions
    e. Tooling maintenance: Is tooling inspected before storing? Is it cleaned and repaired if necessary?
    f. What are the machine shop's capabilities?
    g. How is support equipment handled?
    h. Does the plant use computer control?

6. Quality control
    a. Is quality control part of manufacturing and management philosophy?
    b. To whom do the quality controllers report?
    c. Is a testing area available on site?
    d. How is the testing area equipped?
    e. Are historical samples of parts kept?

7. Other areas to note
    a. How do plant personnel go about solving problems?
    b. Are employees given written work regulations or directions?
    c. Define general shop practices. Are molding problems and their solutions written up? What records are kept on molding conditions and job history?

Some of this information may not be necessary, but if recorded in one place and used to compare one shop with another, it should provide a clear understanding of the type of vendor with whom you are dealing.

PROPRIETARY INFORMATION AGREEMENT

As the developer of a new product, you should take every precaution to see that your new product idea is not stolen from you. Most molders are honest and would not consider passing on proprietary information to anyone. However, it never hurts to have your molder fill out a proprietary information agreement to protect you should information inadvertently get out before you have protected yourself. A sample of such an agreement is given below.

Proprietary Information Agreement

This agreement is made by and entered into between John Q. Public, having a place of business at Seattle, Washington, and XYZ Molding Co. (hereafter "Transferee") having an office and place of business at Seattle, Washington:

Recitals

John Q. Public has certain information relating to apparatus and processes for a product line that may or may not constitute the basis of patentable invention:

John Q. Public considers the above information to be the proprietary information of John Q. Public; and

John Q. Public wishes to transfer and TRANSFEREE wishes to receive the Proprietary Information in connection with considering a proposal to manufacture and supply or otherwise assist in the development of such a product.

Agreements

Accordingly, John Q. Public and TRANSFEREE agree as follows:

1. TRANSFEREE agrees to keep Proprietary Information in confidence and neither make copies nor disclose it to any person or entity not a party to this agreement, not use Proprietary Information except for the purpose set forth above, without the prior written consent of John Q. Public: provided, however, that TRANSFEREE

shall not be liable for use or disclosure of Proprietary Information if it:

a. enters the public domain, prior to such use or disclosure, through no fault of TRANSFEREE;

b. is known to TRANSFEREE, at the time of the transfer, as evidenced by TRANSFEREE's written records; or

c. becomes known to TRANSFEREE, prior to such use or disclosure, without similar restrictions from an independent sources having the right to convey it.

TRANSFEREE will make Proprietary Information available only to its employees having a "need to know" in order to carry out their functions in connection with the product.

TRANSFEREE shall familiarize each such employee having access to Proprietary information with their obligations under this agreement.

2. The transfer of Proptietary Information hereunder shall not be construed as granting TRANSFEREE either a license under and patent or patent application of any license under or right to ownership in said Proprietary Information, nor shall such transfer constitute representation, warranty, assurance, or guarantee to TRANSFEREE with respect to infringement of any rights or others.

3. All Proprietary Information exchanged hereunder shall remain the property of John Q. Public and shall be returned to him or destroyed at his request.

4. This agreement shall be governed by the laws of the state of Washington.

5. The terms of this agreement shall begin on the date it becomes fully executed by both parties, and shall terminate upon 30 days' written notice by either party to such effect. The expiration of this agreement shall not relieve TRANSFEREE of its obligations arising during the term hereof.

XYZ MOLDING COMPANY   JOHN Q. PUBLIC

By: _____   By: _____
Typed name: _____   Typed name: _____
Title: _____   Title: _____
Date: _____   Date: _____

Taking a business-like approach to any project is always a wise procedure. Such things as patents, market studies, venture capital, and proper pricing strategy are often neglected and the project suffers.

PATENTING A PRODUCT*

Often the question comes up; Should I get a patent on my idea, or will that only give it away to someone who will steal it from me? It has become an expensive matter to finance a patent search, hire a patent lawyer, and go through the actions of getting a patent.

Many people say that all a patent does is tell the world about your idea. The only way to keep someone from stealing your idea is to get a head start on any would-be product thief. By building tools and getting the product on the market first, you keep the competition behind you.

What is a patent? That can be answered in simplest form by stating that it is a grant by a government giving an inventor a monopoly on an invention for a fixed period of time. In the language of the U.S. patent laws and the grant itself, a patent is the "right to exclude others from making, using, or selling" the invention the patent covers.

There are five broad rules that should be applied to determine whether an invention is patentable:

1. It must be original and novel.
2. It must be unknown.
3. It must be usable.
4. It must be workable.
5. It must be beyond the idea stage.

Keep accurate records of everything pertaining to your invention. Date all drawings, and if possible, have them witnessed by someone you can trust. Do not throw away any papers pertaining to your invention. Retain all sketches, notes, and drawings, even if they are on the back of a napkin. Carefully retain any correspondence you may have had with anyone about the invention.

According to H. Patrick Thornton, president of Thornton Design, Inc.: "A good invention does not necessarily make a good product.

---

*Material in this section has been taken from Marvin Grosswirth, *How to Patent and Market Your Own Invention*, David McKay Co., Inc., New York, 1978.

An invention is an idea for applying some technology in the service of a specific goal. To make a good product out of a good invention, you must ask more of it. It must be innovative, must satisfy a need, and be attractive to the buyer. It also must be simply and comfortable for the user, cost-competitive, and manufacturable."

THE NEED FOR A MARKETING PLAN

The question often comes up: How much of a formal marketing plan is needed to get a product off the ground? Do I just fly by the seat of my pants? Is it really necessary to write out a formal marketing plan?

The answers to these questions are not the same for everyone. Some situations demand a formal well-written plan of action; others can get by on a few ideas put down in a notebook. The important point is that at least some form of marketing planning is needed for every product. Without some form of plan for taking your product to market, the product is doomed from the beginning.

According to The Conference Board, an independent, nonprofit business research organization: "The formal marketing plan is the symbol and essence of purposeful management. Formulation of a written plan to guide operations of the marketing function...links in a practical way the customer-oriented marketing concept and the principle of management by objectives" [5].

Let's take a look at the routine of a good marketing plan. Keep in mind that there is no "right" plan for everyone. Each person should tailor make a plan to what seems right for him or her. Basically both the ingredients and the purpose are the same. The three main elements of a plan should be:

1. Situation analysis
2. Statement of marketing objectives
3. Summary of strategy and action programs

In determining the market situation, one should look at all possible facts that directly influence the future of the product. David S. Hopkins [2] suggests keeping a "fact book" on the history of a product and its related fields of interest. Such a book would contain a brief report of sales, market share, and competitive trends of similar products. Generally, when you are starting out with a new product, such information may be difficult to obtain. Often with a new product it is hard to find the history of a similar product.

It is in this first section of the report that the present market situation and competitive environment should be spelled out to the best of your ability. Time spent evaluating the competition is time well spent. A detailed review of key competitors' sales, promotional campaigns, product strengths and weaknesses, pricing, and so on, will help immensely in planning a marketing strategy.

In this section you should also spell out any technical costs or other deficiencies that you can foresee. Be objective and honest with yourself. Once you have recognized the problems, it is often easier to plan around them.

The next section of the report should center around market objectives. Just what is it that you want to accomplish? It could be to capture as much of the market as is available to you. Or perhaps it is a set profit figure. The one thing to remember is to be realistic and define your objectives clearly.

Finally, you should outline your plan of action. Detail your plans so that there is no misunderstanding of what it is you plan to do to accomplish your goals. A list of actions to be taken might include:

1. Developing a prototype of your product
2. Running a test market of your product
3. Establishing a realistic pricing program
4. Developing a packaging program for your product
5. Developing sales channels to get your product to market

Together with these action plans should go a timetable that you can work with. Be reasonable about the time frame and give yourself enough time to get the job done.

Once you have written this marketing plan, use it. Refer to it often and keep changing it as the situation changes. If you follow this plan, you will have a much better chance of succeeding in the development of your product.

PRICING A PRODUCT*

One of the most important problems facing a project manager is how to price a product properly. Many books have been written about this subject and you may want to consult these to investigate the subject further.

As an overall objective, you want to establish a price that will result both in securing orders for the product and providing a reasonable profit. Accomplishing this objective requires that you do several things:

1. Evaluate the actual cost to manufacture the product.
2. Establish a price that takes into consideration the following:

   a. Competition
   b. Current and future workloads
   c. Market strategy
   d. Profit objectives

To determine the actual cost of the product, it is necessary to collect the most accurate and reliable information available. This includes the following:

1. The cost of all material to be used in the product. Do not neglect anything, from packaging to the smallest fastener.
2. The next item to consider is tooling costs. Since this usually represents the largest initial investment, you should get an accurate idea of what these costs will be. Do not forget the cost of any jig or assembly fixtures that will be needed in the manufacture of the item.
3. The time needed to produce the item and the type and number of personnel are next. This is often the hardest figure to estimate. If you can use a similar product as an example, this will often give you an idea of the time required.
4. Labor costs must be established. In a labor-intensive product, this cost can amount to a large share of the overall costs.
5. Finally, estimate the overall sales and general administration costs.

Once you have gathered all this information on the product, you can establish a reasonably good cost of producing your product. Some other items to take into consideration are the impact of inflation on the cost estimate and any scrap factors that may influence the cost figures.

With the cost factors in mind, it is time to establish profit objectives. Two ways of looking at profit are return on sales and return on investment. You should plan on making enough profit to provide a good return to the owners of the company and provide for expansion or modernization. Before setting a profit objective, you should also consider what the general business environment will be next year.

Several basic assumptions must be made in your pricing considerations:

1. The cost/selling price ratio will vary over the life of a product.
2. Pricing is influenced by capacity-utilization considerations.
3. Prices are sensitive to the profit strategies of those companies that use the product.
4. Prices are sensitive to competition.

Therefore, an established price should be based on the competitive environment, the company's market strategy regarding the product, current and future workloads, the company's profit objective, and the estimated cost.

Let's look at several ways to establish price. First is the *return-to-investment* technique. Proceed as follows:

1. Determine the desired pretax income.
2. Establish selling, general, and administrative expenses.
3. Add the above to determine the required return on inventory and fixed assets. This return is the required gross profit.
4. Estimate material, direct labor, and manufacturing overhead costs.
5. Identify and value the total assets to be included in the computation.
6. Compute the return required as a percentage of the total assets being considered.
7. Determine the turnover of the assets in relation to the estimated costs.
8. Compute the price mark-up to be used by dividing the turnover factor into the percentage return. These mark-on percentages are applied to their respective costs to establish a price.

Study the following example.

	Estimated cost (1)	Investment (2)	Investment turnover (3) = (1)/(2)
Material	$3,860,000	$ 950,000	4.0
Conversion cost (direct labor and manufacturing overhead)	2,380,000	1,500,000	1.6
Manufacturing cost	$6,240,000	$2,450,000	2.53
	Percent overall return (4)	Markup percent (5) = (4)/(3)	Gross profit (6) = (1) × (5)
Material	70.2%	17.6%	$ 679,360
Conversion cost (direct labor and manufacturing overhead)	70.2%	43.9%	1,044,820
Manufacturing cost	70.2%	27.7%	$1,724,180

Markup applied to manufacturing costs:
Manufacturing cost: ($34.98); markup: (27.7%) = 9.69.
Target price: $34.98 + 9.69 = $44.67

Other pricing techniques exist and can be utilized. It is important to have a pricing plan and to stick to it.

VENTURE CAPITAL

Often in the development of a plastics project, the subject of venture capital is raised. Too often there is just not enough money available to get a project off the ground and an otherwise excellent idea goes down to failure. It is not just having enough funds to get the tooling started—one must have enough capital to exist during the development period. One way of providing such funds is through venture capital groups.

"I'll back you if you have a good idea that will make money for both of us [6]. This sentence says it all. To attract venture capital, one must be prepared to put the *entire* plan in proposal form. That is, lay out the project so that a venture capitalist can have a complete idea of the scope of the project and thus can make an entirely informed decision. Venture capitalists, on average, get 50 to 100 telephone calls a day asking for capital. It is clear that a proposal must be complete and very interesting to warrant attention.

The basic outline of a proposal can vary, but it should contain at least these items:

1. A summary
2. The business and its future
3. The management personnel
4. A description of the financing
5. The risk factors
6. Return on investment and the way a venture capitalist could exit the venture
7. An analysis of operations and projections
8. Financial statements
9. Projections
10. Illustrative information

More information on this subject is available in the many books that have been written on obtaining venture capital.

Where can you find venture capital? Securing a list of names is relatively simply. Two published lists developed by the following organizations are available for the asking.

National Association of Small Business Investment
618 Washington Building
Washington, DC 20005
(202) 638-3411

The National Venture Capital Association
1225 19th Street NW, Suite 750
Washington, DC 20036
(202) 659-5756

VERTICAL INTEGRATION

Often in the development of any type of plastics project, the question arises: Should we mold the parts ourselves? This question can only be answered by looking at the cost figures involved in going into the molding business and by looking at the anticipated volumes that you will be producing in the future.

For the purposes of developing some cost figures, we look at a typical molding operation and some of the expenses needed to set up a profitable operation. We then compare these costs with the projected volumes and see what we need to justify going into such an operation.

We look first at the equipment required and the cost to install it.

Machinery
Injection molding machine (175 tons, 6 oz)	$175,000
Chiller	6,000
Material handling system	5,000
Dryer	6,000
Grinder	4,000
Forklift truck (used)	6,000
Installation charges and freight	7,000
Total capital costs to set up operation	$209,000

Other costs
Labor (2 workers/year—3 shifts)	$108,000
Utilities	18,000
Rent	35,000
Total major costs	$437,000

To put this together represents a substantial outlay of funds and means a high cost of production unless you have enough work to keep the machine in operation on 5 three-shift days out of 7. Injection molding machines are designed to be run 24 hours a day. A project manager should be aware of the high cost of starting a molding operation and be prepared to have enough parts to keep such an operation busy before recommending such a move.

REFERENCES

1. Mel Friedman, editorial, *Plastics Design Forum*, Jan.–Feb.
2. *Plastics Design Manual*, Borg-Warner Chemicals, Chicago.
3. Geoffrey Boothroyd and Peter Dewhurst, *Machine Design*, Jan. 26, 1984.
4. Geoffrey Boothroyd and Peter Dewhurst, *Design for Assembly Handbook*, University of Massachusetts, Boston, 19XX.
5. David S. Hopkins, *The Short Term Marketing Plan*, The Conference Board, 1972.
6. David J. Gladstone, *Venture Capital Handbook*, Reston Publishing Co., Inc., Reston, Va., 1983.

Appendix I
Selected Conversion Factors

To convert from	to:	multiply by:
atmosphere (760 mm Hg)	pascal (Pa)	$1.013\ 25 \times 10^5$
board foot	cubic meter (m^3)	$2.359\ 737 \times 10^{-3}$
Btu (International Table)	joule (J)	$1.055\ 056 \times 10^3$
Btu (International Table)/h	watt (W)	$2930\ 711 \times 10^{-1}$
Btu (International Table)·in./s·ft²·°F (k, thermal conductivity)	watt per meter kelvin [W/(m·K)]	$5.192\ 204 \times 10^2$
calorie (International Table)	joule (J)	$4.186\ 800*$
centipoise	pascal second (Pa·s)	$1.000\ 000* \times 10^{-3}$
centistoke	square meter per second (m^2/s)	$1.000\ 000*\ 10^{-3}$
circular mil	square meter (m^2)	$5.067\ 075 \times 10^{-10}$
degree Fahrenheit	degree Celsius	$t_{°C} = (t_{°F} - 32)/1.8$
foot	meter (m)	$3.048\ 000* \times 10^{-1}$
ft²	square meter (m^2)	$9.290\ 304*\ 10^{-2}$
ft³	cubic meter (m^3)	$2.831\ 685 \times 10^{-2}$
ft·lbf	joule (J)	$1.355\ 818$
ft·lbf/min	watt (W)	$2.259\ 697 \times 10^{-2}$

(continued)

To convert from	to:	multiply by:
ft/s^2	meter per second squared (m/s^2)	$3.048\ 000* \times 10^{-1}$
gallon (U.S. liquid)	cubic meter (m^3)	$3.785\ 412 \times 10^{-3}$
horsepower (electric)	watt (W)	$7.460\ 000* \times 10^{-2}$
inch	meter (m)	$2.540\ 000* \times 10^{-2}$
in.^{2}	square meter (m^2)	$6.451\ 600* \times 10^{-4}$
in.^{3}	cubic meter (m^3)	$1.638\ 706 \times 10^{-5}$
inch of mercury (60 °F)	pascal (Pa)	$3.376\ 85 \times 10^3$
inch of water (60 °F)	pascal (Pa)	$2.488\ 4 \times 10^2$
kgf/cm^2	pascal (Pa)	$9.806\ 650* \times 10^4$
kip (100 lbf)	newton (N)	$4.448\ 222 \times 10^3$
kip/in.^{2} (ksi)	pascal (Pa)	$6.894\ 757\ ts\ 10^6$
ounce (U.S. fluid)	cubic meter (m^3)	$2.957\ 353 \times 10^{-5}$
ounce-force	newton (N)	$2.780\ 139 \times 1^{-1}$
ounce (avoirdupois)	kilogram (kg)	$2.834\ 952 \times 10^{-2}$
oz (avoirdupois)/ft^2	kilogram per square meter (kg/m^2)	$3.051\ 517 \times 10^{-1}$
oz (avoirdupois)/ yd^2	kilogram per square meter (kg/m^2)	$3.390\ 575 \times 10^{-2}$
oz (avoirdupois)/ gal (U.S. liquid)	kilogram per cubic meter (kg/m^3)	$7.489\ 152$
pint (U.S. liquid)	cubic meter (m^3)	$4.731\ 765 \times 10^{-4}$
pound-force (lbf)	newton (N)	$4.448\ 222$
pound (lb avoirdupois)	kilogram (kg)	$4.535\ 924 \times 10^{-1}$
lbf/in^2 (psi)	pascal (Pa)	$6.894\ 757 \times 10^3$
lb/in.^{3}	kilogram per cubic meter (kg/m^3)	$2.767\ 990 \times 10^4$
lb/ft^3	kilogra per cubic meter (kg/m^3)	$1.601\ 846 \times 10$
quart (U.S. liquid)	cubic meter (m^3)	$9.463\ 529 \times 10^{-4}$
ton (short, 2000 lb)	kilogram (kg)	$9.071\ 847 \times 10^2$
torr (mm Hg, 0 °C)	pascal (Pa)	$1.333\ 22 \times 10^2$
W·h	joule (J)	$3.600\ 000* \times 10^3$
yard	meter (m)	$9.144\ 000* \times 10^{-1}$
yd^2	square meter (m^2)	$8.361\ 274 \times 10^{-1}$
yd^3	cubic meter (m^3)	$7.645\ 549 \times 10^{-1}$

Source: ASTM F380.
*Exact.

Appendix II
History of Plastics

In any book on plastics it is important to review the history of the material and its many applications that have benefited humankind. Such a review should provide a better understanding of the sequence of introduction of various applications.

1868

- John Wesley Hyatt mixed proxylin and nitric acid with camphor to create cellulose nitrate. Named Celluloid, it was recognized as the first commercial plastic in the United States and was developed as a substitute material for ivory in billiard balls.

1909

- The development of Bakelite phenolic is accomplished by Leo Hendrick Baekeland.

Material in this appendix has been taken from *50 Years of Progress in Plastics*. HBJ Plastics Publication 22–41, Society of Plastics Industry, 1987.

1937

- In Leverkusen, Germany, Otto Bayer develops the diisocyanate polyurethane addition process, which will act as the basis of polyurethane chemistry.
- The Hudson Car Co. uses plastic in its radiator ornaments, electrical parts and in the first fiber-impregnated gear.

1938

- Plax Corp. purchases a fully automatic thermoforming machine from Clauss B. Slauch for the production of cellulose acetate, Christmas-tree spires, cigarette premiums, and ice-cube trays.
- W. K. Kopitke invents the injection blow molding process (patent issued in 1943).
- Du Pont announces Wallace H. Carothers' invention of nylon.
- Reed Prentice introduces the 10-D-8 injection machine (which sold over 2000 units).

1939

- Hercules begins production of cellulose acetate molding compound.

1940

- Bomber noses of Plexiglass acrylic resin are thermoformed for war planes.
- John Reilly and Ralph Wiley of Dow Chemical introduce vinylidene chloride in film and sheeting.
- L. C. Kiesel Co. Designs an electric guitar that is compression molded of Resinox by F. E. Reinhold for Continental Music Co.

1941

- Kirk Molding Co. receives a contract from Du Pont's Doyle Works for molding combs and toothbrush handles.

1942

- Polyethylene is introduced.

- Nash Motors selects polystyrene for its horn medallion, a showpiece part.
- The first polyethelene bottle is blown with technology developed by James T. Bailey of Plax Corp. (later discarded for lack of clarity and rigidity).
- Becton Dickinson develops the thermoformed blister package.

1943

- Saran monofilament is extruded by Dow Chemical.
- Hercules Powder Co. supplies cellulose acetate for plastic toothpaste tubes.
- Dow Chemical begins production of silicone resins.

1944

- Plax salesmen are given samples of polyethylene squeeze bottles.

1945

- W. Brandt Goldsworthy molds the first all-fiberglass-reinforced automobile for Consolidated Aircraft at his Industrial Plastics Co.; four were built.

1946

- Chrysler introduces acrylic automobile taillight lenses.
- Jim Hendry builds a 2-oz screw injection machine.
- Waldes Kohinoor Inc. introduces the nylon zipper.

1948

- Admiral 35-lb TV cabinet is molded of walnut and mahogany-colored phenolic.

1950

- Hermam Miller introduces the fiberglass-reinforced plastic "shell chair" designed by Charles Eames.

1952

- William H. Willert invents an in-line reciprocating screw plasticating injection molding machine.
- B.F. Goodrich develops high-impact rigid PVC, which makes possible the development of plastic pipe.
- Molded Fiber Glass Tray Co. molds glass-reinforced plastic bread delivery trays for Wonder Bread, the first reinforced-plastic product (more than 400,000 units were produced).

1954

- Ford's Thunderbird uses a reinforced-plastic hardtop.

1955

- General Electric introduces polycarbonate.

1956

- Lucent's Melamine dinnerware brings "new" dignity and a larger market to this type of product.

1957

- The Hoola-Hoop boom taxes the capacity of the extrusion industry and the capacity for high-density polyethelene at a peak use of 1 million pounds per week.
- The 1957 Ford Thunderbird inaugurates widespread use of urethane foam for seating and safety dashboards.
- Loma Industries molds the first 20-gallon trash containers.

1958

- Monsanto blow-molds the first experimental Coca-Cola bottle from acrylonitrile.

1959

- Du Pont begins commercial production of acetal homopolymer.
- Mattel Inc. introduces the record-breaking plastic Barbie Doll.

1962

- Polymers, introduced by Du Pont, increase the thermal endurance of thermoplastics to 750 °F.
- HeyWoodite school furniture combines solid and laminated plastic in a tubular steel construction.

1963

- Richard G. Angell Jr. invents the Union Carbide process for low-pressure structural foam.
- Frank H. Lambert develops expanded polystyrene foam products for molding.
- The Studebaker Avanti, designed by Raymond Loewy, demonstrates the first all-reinforced-plastic body on a hardtop sedan.

1965

Alfred Farnham and Robert Johnson develop polysulfone, introduced by Union Carbide.

1966

The first Uniloy machine for the production on HDPE blow-molded milk bottles is installed at the Heatherwood Diary.
- General Electric introduces modified polyphenylene oxide.
- Thermoplastics show up in exterior auto applications as Pontiac uses nylon for fender extensions.

1972

- Martin Industries receives a patent on its cam-actuated grabber, the first high-speed machine-mounted automatic part remover for injection molding.

1973

- Energy-absorbing microcellular urethane foam, a forerunner to RIM, is tested on a New York taxi fleet and appears on the Chevelle Laguna.

1974

- The first RIM fascia debuts on the Pontiac line.

1975

- RIM urethane comes of age in the Chevrolet Monza front end, which is held in place by a retainer of glass-reinforced polypropylene sheeting.

1976

- Plastic microwave cookware enters the consumer market.

1977

- The 2-liter oriented PET bottle goes into commercial production.

1979

- Volume production of plastics surpasses that of steel.
- Apple Computer produces its Apple IIE personal computer in ABS engineering plastic at a cost savings and a gain in surface appearance, flame retardance, and heat distortion characteristics.

1982

- Bayer AG researchers, working in Europe with the developers of compact audio disks, introduce high-purity polycarbonate resin for the manufacture of CDs.
- Robert Jarvik designs the Jarvik heart, made largely of plastics, which supports Barney Clark's life for 112 days.

1983

- FCC regulations require the shielding of plastic-housed electronic components, which spurs the development of conductive composites for computer housings.
- American Can Packaging, Inc. produces a 28-oz squeezable plastic bottle for Heinz catsup.

1984

- Five sets of composite horizontal stabilizers are installed on the Boeing 737.
- Wilson's squeeze test can enables tennis players to check tennis balls before purchase.
- The first plastic fuel tank in a U.S. passenger car is blow-molded of Phillip's HDPE by Bronson.
- Quaker State moves to HDPE bottles for its entire motor-oil line.
- Herman Miller introduces the Equa chair.

These and hundreds of similar developments over the past century have put plastics in the forefront of the technological expansion of our civilization.

Appendix III
Selected Plastics Suppliers

Allied	(201) 455-5010
Amoco	800-621-8888
Arco	800-354-1440
Atochem Inc.	800-932-0420
BASF	800-521-9100
Borg-Warner	(304) 424-5411
Chevron	(713) 754-4243
Dow	800-258-2436
Du Pont	800-828-6876 or 800-441-7515
Emser Industries	800-481-3172
Firestone	800-282-0222
General Electric	800-845-0600
Himont	800-247-4372
Hoechst Celanese	800-CELANESE
Mitsui	(212) 876-4462
Mobay	800-345-8345
Mobile	(203) 629-8810
Monsanto	800-325-4330
Phillips	800-53RESIN
Shell	800-323-3405
Wilson Fiberfill	800-457-3764

Appendix IV

STANDARDS AND PRACTICES OF PLASTICS MOLDERS	Engineering and Technical Standards POLYCARBONATE

NOTE: The Commercial values shown below represent common production tolerances at the most economical level. The Fine values represent closer tolerances that can be held but at a greater cost.

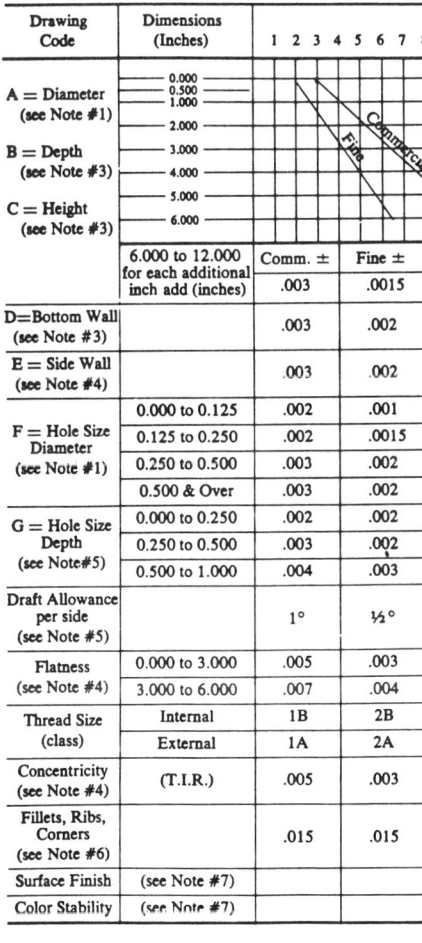

Drawing Code	Dimensions (Inches)	Plus or Minus in Thousands of an Inch 1 2 3 4 5 6 7 8 9 10 11 12 13 14 15 16 17 18 19 20 21 22 23 24 25 26 27 28
A = Diameter (see Note #1)	0.000 / 0.500 / 1.000 / 2.000	
B = Depth (see Note #3)	3.000 / 4.000	
C = Height (see Note #3)	5.000 / 6.000	

Drawing Code	Dimensions (Inches)	Comm. ±	Fine ±
	6.000 to 12.000 for each additional inch add (inches)	.003	.0015
D=Bottom Wall (see Note #3)		.003	.002
E = Side Wall (see Note #4)		.003	.002
F = Hole Size Diameter (see Note #1)	0.000 to 0.125	.002	.001
	0.125 to 0.250	.002	.0015
	0.250 to 0.500	.003	.002
	0.500 & Over	.003	.002
G = Hole Size Depth (see Note#5)	0.000 to 0.250	.002	.002
	0.250 to 0.500	.003	.002
	0.500 to 1.000	.004	.003
Draft Allowance per side (see Note #5)		1°	½°
Flatness (see Note #4)	0.000 to 3.000	.005	.003
	3.000 to 6.000	.007	.004
Thread Size (class)	Internal	1B	2B
	External	1A	2A
Concentricity (see Note #4)	(T.I.R.)	.005	.003
Fillets, Ribs, Corners (see Note #6)		.015	.015
Surface Finish	(see Note #7)		
Color Stability	(see Note #7)		

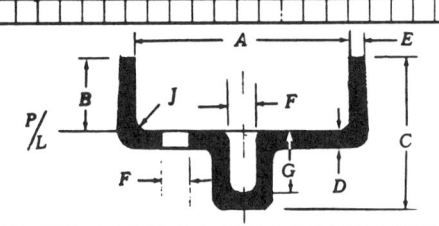

REFERENCE NOTES

1 — These tolerances do not include allowance for aging characteristics of material.

2 — Tolerances based on ⅛" wall section.

3 — Parting line must be taken into consideration.

4 — Part design should maintain a wall thickness as nearly constant as possible. Complete uniformity in this dimension is impossible to achieve.

5 — Care must be taken that the ratio of the depth of a cored hole to its diameter does not reach a point that will result in excessive pin damage.

6 — These values should be increased whenever compatible with desired design and good molding technique.

7 — Customer-Molder understanding necessary prior to tooling.

STANDARDS AND PRACTICES OF PLASTICS MOLDERS	Engineering and Technical Standards ABS

NOTE: The Commercial values shown below represent common production tolerances at the most economical level. The Fine values represent closer tolerances that can be held but at a greater cost.

Drawing Code	Dimensions (Inches)			Plus or Minus in Thousands of an Inch 1 2 3 4 5 6 7 8 9 10 11 12 13 14 15 16 17 18 19 20 21 22 23 24 25 26 27 28
A = Diameter (see Note #1) B = Depth (see Note #3) C = Height (see Note #3)	0.000 0.500 1.000 2.000 3.000 4.000 5.000 6.000			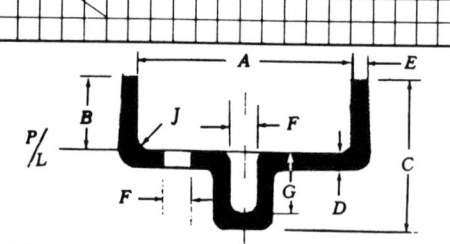

Drawing Code	Dimensions (Inches)	Comm. ±	Fine ±
	6.000 to 12.000 for each additional inch add (inches)	.003	.002
D=Bottom Wall (see Note #3)		.004	.002
E = Side Wall (see Note #4)		.003	.002
F = Hole Size Diameter (see Note #1)	0.000 to 0.125	.002	.001
	0.125 to 0.250	.002	.001
	0.250 to 0.500	.003	.002
	0.500 & Over	.004	.002
G = Hole Size Depth (see Note#5)	0.000 to 0.250	.003	.002
	0.250 to 0.500	.004	.002
	0.500 to 1.000	.005	.003
Draft Allowance per side (see Note #5)		2°	1°
Flatness (see Note #4)	0.000 to 3.000	.015	.010
	3.000 to 6.000	.030	.020
Thread Size (class)	Internal	1	2
	External	1	2
Concentricity (see Note #4)	(T.I.R.)	.009	.005
Fillets, Ribs, Corners (see Note #6)		.025	.015
Surface Finish	(see Note #7)		
Color Stability	(see Note #7)		

REFERENCE NOTES

1 – These tolerances do not include allowance for aging characteristics of material.

2 – Tolerances based on ⅛″ wall section.

3 – Parting line must be taken into consideration.

4 – Part design should maintain a wall thickness as nearly constant as possible. Complete uniformity in this dimension is impossible to achieve.

5 – Care must be taken that the ratio of the depth of a cored hole to its diameter does not reach a point that will result in excessive pin damage.

6 – These values should be increased whenever compatible with desired design and good molding technique.

7 – Customer-Molder understanding necessary prior to tooling.

STANDARDS AND PRACTICES OF PLASTICS MOLDERS

Engineering and Technical Standards
POLYPROPYLENE

NOTE: The Commercial values shown below represent common production tolerances at the most economical level. The Fine values represent closer tolerances that can be held but at a greater cost.

Drawing Code	Dimensions (Inches)	Plus or Minus in Thousands of an Inch
A = Diameter (see Note #1)	0.000 / 0.500 / 1.000 / 2.000	1 2 3 4 5 6 7 8 9 10 11 12 13 14 15 16 17 18 19 20 21 22 23 24 25 26 27 28
B = Depth (see Note #3)	3.000 / 4.000	
C = Height (see Note #3)	5.000 / 6.000	

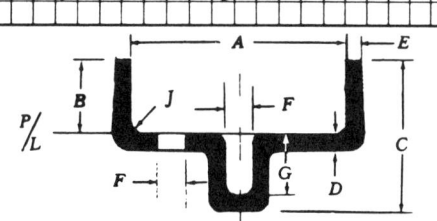

		Comm. ±	Fine ±
	6.000 to 12.000 for each additional inch add (inches)	.005	.003
D=Bottom Wall (see Note #3)		.006	.003
E = Side Wall (see Note #4)		.006	.003
F = Hole Size Diameter (see Note #1)	0.000 to 0.125	.003	.002
	0.125 to 0.250	.004	.003
	0.250 to 0.500	.005	.004
	0.500 & Over	.008	.006
G = Hole Size Depth (see Note#5)	0.000 to 0.250	.005	.003
	0.250 to 0.500	.006	.004
	0.500 to 1.000	.009	.006
Draft Allowance per side (see Note #5)		1½°	½°
Flatness (see Note #4)	0.000 to 3.000	.021	.014
	3.000 to 6.000	.035	.021
Thread Size (class)	Internal	1	2
	External	1	2
Concentricity (see Note #4)	(T.I.R.)	.016	.013
Fillets, Ribs, Corners (see Note #6)		.028	.015
Surface Finish	(see Note #7)		
Color Stability	(see Note #7)		

REFERENCE NOTES

1 – These tolerances do not include allowance for aging characteristics of material.

2 – Tolerances based on ⅛″ wall section.

3 – Parting line must be taken into consideration.

4 – Part design should maintain a wall thickness as nearly constant as possible. Complete uniformity in this dimension is impossible to achieve.

5 – Care must be taken that the ratio of the depth of a cored hole to its diameter does not reach a point that will result in excessive pin damage.

6 – These values should be increased whenever compatible with desired design and good molding technique.

7 – Customer-Molder understanding necessary prior to tooling.

STANDARDS AND PRACTICES OF PLASTICS MOLDERS	Engineering and Technical Standards POLYSTYRENE

NOTE: The Commercial values shown below represent common production tolerances at the most economical level. The Fine values represent closer tolerances that can be held but at a greater cost.

Drawing Code	Dimensions (Inches)	Plus or Minus in Thousands of an Inch 1 2 3 4 5 6 7 8 9 10 11 12 13 14 15 16 17 18 19 20 21 22 23 24 25 26 27 28
A = Diameter (see Note #1) B = Depth (see Note #3) C = Height (see Note #3)	0.000 0.500 1.000 2.000 3.000 4.000 5.000 6.000	*Commercial / Fine graph*

	6.000 to 12.000 for each additional inch add (inches)	Comm. ± .004	Fine ± .002
D=Bottom Wall (see Note #3)	•	.0055	.003
E = Side Wall (see Note #4)		.007	.0035
F = Hole Size Diameter (see Note #1)	0.000 to 0.125	.002	.001
	0.125 to 0.250	.002	.001
	0.250 to 0.500	.002	.0015
	0.500 & Over	.0035	.002
G = Hole Size Depth (see Note#5)	0.000 to 0.250	.0035	.002
	0.250 to 0.500	.004	.002
	0.500 to 1.000	.005	.003
Draft Allowance per side (see Note #5)		1½°	½°
Flatness (see Note #4)	0.000 to 3.000	.007	.004
	3.000 to 6.000	.013	.005
Thread Size (class)	Internal	1	2
	External	1	2
Concentricity (see Note #4)	(T.I.R.)	.010	.008
Fillets, Ribs, Corners (see Note #6)		.015	.010
Surface Finish	(see Note #7)		
Color Stability	(see Note #7)		

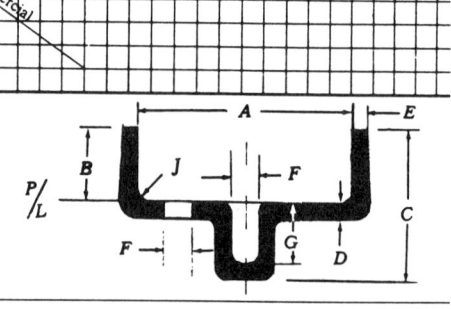

REFERENCE NOTES

1 — These tolerances do not include allowance for aging characteristics of material.

2 — Tolerances based on ⅛″ wall section.

3 — Parting line must be taken into consideration.

4 — Part design should maintain a wall thickness as nearly constant as possible. Complete uniformity in this dimension is impossible to achieve.

5 — Care must be taken that the ratio of the depth of a cored hole to its diameter does not reach a point that will result in excessive pin damage.

6 — These values should be increased whenever compatible with desired design and good molding technique.

7 — Customer-Molder understanding necessary prior to tooling.

STANDARDS AND PRACTICES OF PLASTICS MOLDERS

Engineering and Technical Standards
HIGH DENSITY POLYETHYLENE

NOTE: The Commercial values shown below represent common production tolerances at the most economical level. The Fine values represent closer tolerances that can be held but at a greater cost.

Drawing Code	Dimensions (Inches)		Comm. ±	Fine ±
A = Diameter (see Note #1)	0.000 / 0.500 / 1.000 / 2.000			
B = Depth (see Note #3)	3.000			
C = Height (see Note #3)	4.000 / 5.000 / 6.000			
	6.000 to 12.000 for each additional inch add (inches)	Comm. ±	.006	Fine ± .003
D = Bottom Wall (see Note #3)			.006	.004
E = Side Wall (see Note #4)			.006	.004
F = Hole Size Diameter (see Note #1)	0.000 to 0.125		.003	.002
	0.125 to 0.250		.005	.003
	0.250 to 0.500		.006	.004
	0.500 & Over		.008	.005
G = Hole Size Depth (see Note#5)	0.000 to 0.250		.005	.003
	0.250 to 0.500		.007	.004
	0.500 to 1.000		.009	.006
Draft Allowance per side (see Note #5)			2°	¾°
Flatness (see Note #4)	0.000 to 3.000		.023	.015
	3.000 to 6.000		.037	.022
Thread Size (class)	Internal		1	2
	External		1	2
Concentricity (see Note #4)	(T.I.R.)		.027	.010
Fillets, Ribs, Corners (see Note #6)			.025	.010
Surface Finish	(see Note #7)			
Color Stability	(see Note #7)			

Plus or Minus in Thousands of an Inch
1 2 3 4 5 6 7 8 9 10 11 12 13 14 15 16 17 18 19 20 21 22 23 24 25 26 27 28

Commercial line

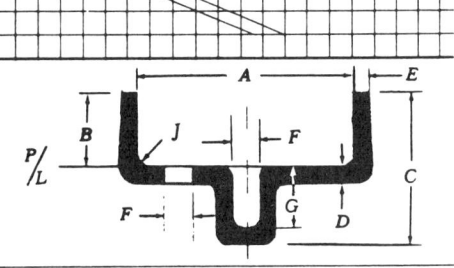

REFERENCE NOTES

1 — These tolerances do not include allowance for aging characteristics of material.

2 — Tolerances based on ⅛" wall section.

3 — Parting line must be taken into consideration.

4 — Part design should maintain a wall thickness as nearly constant as possible. Complete uniformity in this dimension is impossible to achieve.

5 — Care must be taken that the ratio of the depth of a cored hole to its diameter does not reach a point that will result in excessive pin damage.

6 — These values should be increased whenever compatible with desired design and good molding technique.

7 — Customer-Molder understanding necessary prior to tooling.

Bibliography

Adhesives in Manufacturing, F.L. Schneberger Publications, 1983.

Coloring of Plastics, Thomas G. Wevver, Society of Plastics, Engineers, Brookfield Center, CT, 1979.

Composites: A Design Guide, Terry Richardson, Society of Plastics, Brookfield Center, CT, 1987.

Cost Estimating and Pricing, Robert Gorski, Dwight L. Hangar, and Terrance R. Orzan, Financial Management Committee, The Society of Plastics Industry, Inc.,

Decorating Plastics, James M. Margolis, Society of Plastics Engineers, Brookfield Center, CT, 1986

Designing with Plastics, G. W. Ehrenstein, Society of Plastics Engineers, Brookfield Center, CT, 1984.

Engineering Guide to Plant Layout and Machine Selection, Bruce C. Wendle, Technomic Publishing Co., Inc., Lancaster, PA, 19XX.

Engineering Guide to Structural Foam, Bruce C. Wendle, Technomic Publishing Co., Inc., Lancaster, PA, 1976.

Engineering Thermoplastic Properties and Applications, James M. Margolis, Society of Plastics, Engineers, Brookfield Center, CT, 1985.

How to Patent and Market Your Own Invention, Marvin Grosswirth, David McKay Company, Inc., New York, 1978.

Introduction to Extrusion, P. N. Richardson, Society of Plastics Engineers, Brookfield Center, CT, 19XX.

Marketing Problem Solver, Cochrane Chase and Kenneth L. Barasch, Cilton Book Company, Radnor, PA, 1973.

Plastic Product Design, 2nd ed., Ronald D. Beck, Van Nostrand Reinhold Co., Inc., New York, 1980.

Plastics Common Objects, Sylvia Katz, T/C Publications, 1984.

Plastics, Vol. 9, Thermoplastics and Thermosets, Society of Plastics Engineers, Brookfield Center, CT, 1987.

Plastics Finishing and Decoration, Don Satas, Society of Plastics Engineers, Brookfield Center, CT, 1986.

Pricing Strategies, Alfred R. Oxenfeldt, Amacom Book Division, New York, 1975.

Quality Control for Plastics, William J. Tobin, Society of Plastics Engineers, Brookfield Center, CT, 1986.

Short Term Marketing Plan, David S. Hopkins, The Conference Board, 1972.

Structural Design with Plastics, 2nd ed., B.S. Benjamin, Society of Plastics Engineers, Brookfield Center, CT, 1982.

Structural Foam: A Purchasing and Design Guide, Bruce C. Wendle, Marcel Dekker, Inc., New York, 1985.

Thermoforming, James L. Throne, Society of Plastics Engineers, Brookfield Center, CT, 1987.

Thermoplastic Elastomers, N. R. Legge, Society of Plastics Engineers, Brookfield Center, CT, 1987.

Venture Capital Handbook, David J. Gladstone, Reston Publishing Co., Inc., Reston, Va., 1983.

Index

ABS (acrylonitrile, butediene, styrene), 42, 48, 52, 53, 54, 55, 56, 57, 59, 67, 101, 168, 174

Acetal, 43, 48, 52, 53, 54, 56, 57, 59, 61

Acrylic, 43, 48, 52, 53, 55, 56, 59, 62, 101, 164

American Society for Testing & Materials, 20, 58

Arc resistance, 58, 73, 78, 95

Assembly, 73, 78, 95, 141

Blowing agents, 73, 78, 95

Blow molding, 78

Bonding, 105

Braiding, 81

Brainstorming, 18

Cellular products, 32

Coinjection, 76

Color concentrates, 66

Compression molding, 164

Composites, 1, 79

Computer-aided design, 18, 30

Conversion factors, 161, 162

Cost estimating, 89, 90

Crystalline polymers, 5

Dielectric constant, 58

Dielectric strength, 47

Draft, 27, 29

Dry color, 66

Du Pont, 7

Dyeing, 66

Electrical properties, 59

183

Expanded polystyrene foam, 82

Extrusion, 82, 83

Federal Aviation Administrtion, 6, 9

Finishing, 70, 71

First article inspection, 36

General Electric, 7, 21, 30, 40

Hand lay-up, 80

History of plastics, 163, 164, 165, 166

Injection molding, 76, 77, 85, 86, 158, 164, 165, 167

Inserts, 35

In-situ polymerization, 13

Izod impact testing, 39

Just-in-time delivery, 36

Let-down ratio, 66

Liquid polymer, 5, 53, 56, 57, 80

Marketing plan, 152

Material selection, 29, 35, 39, 40, 69

Mold flow, 10, 11

Mold shrinkage, 34, 58, 63

Moments of inertia, 20, 46

Nylon, 43, 49, 52, 53, 54, 55, 56, 57, 59, 60

Packaging, 35, 37

Painting of plastics, 67, 70, 71, 101

Patents, 151

Polyamide imide, 6, 44, 53, 54, 56, 57, 80

Polyarylate, 44, 49, 53, 54, 55

Polyarysulfone, 44, 49, 53, 55, 59

Polyester, 10, 44, 49, 53, 54, 56, 62

Polycarbonate, 43, 49, 52, 54, 55, 56, 57, 59, 62, 101, 173

Polyether ether ketone, 6, 44, 49, 53, 54, 56, 80

Polyether imide, 6, 44, 49, 50, 53, 54, 56, 57, 59

Polyethylene, 10, 41, 50, 52, 54, 55, 57, 101, 177

Polypropylene, 42, 50, 51, 52, 53, 54, 55, 57, 59, 60, 101, 175

Polyphenylene oxide, 43, 50, 52, 54, 55, 59, 101, 167

Polyphenylene sulfide, 44, 50, 53, 54, 57, 59, 80

Polystyrene, 32, 42, 51, 52, 54, 55, 56, 57, 60, 176

Polysulfone, 44, 49, 53, 54,
   56, 80
Polyurethane, 77, 164, 166
Polyvinyl chloride, 10, 51, 52,
   53, 55, 57, 59, 62, 101
Pressure vessels, 24, 26, 27
Pricing, 153, 154, 155
Product design, 133, 138, 139,
   142
Proprietary information, 149
Prospective vendor, 146, 147
Prototyping, 144, 153
Pultrusion, 81

Quality control, 36, 72, 73,
   148

Reaction injection molding,
   13, 32, 43, 77, 93, 92, 167,
   168
Resin transfer, 81
Roto-molding, 75, 91, 92

Safety factors, 26
Sheet molding, 76
Spray-up, 80
Structural foam, 32, 75, 76,
   94

Technical molding concepts,
   12
Test marketing, 35
Thermoplastic, 2, 4, 5, 53,
   77, 80
Thermoplastic elastomer, 44
Thermoset, 4, 80
Tolerances, 135, 136, 173,
   174, 175, 176, 177
Tooling, 91, 93, 97, 101
Troubleshooting injecting
   molding, 86, 87, 88
Twin sheet forming, 76, 101

Ultraviolet light degradation,
   68, 69, 140, 142
Underwriters laboratories, 6, 9
Unisys, Inc. 10, 12

Vacuum bag molding, 81
Vacuum forming, 75, 76, 97,
   100, 101
Venture capital, 156, 157
Vertical integration, 158
Volumn resistivity, 58

Warpage, 60

About the Author

Bruce C. Wendle is a Specialist Engineer at Boeing Commerical Airplane Group, Seattle, Washington. The author of numerous technical articles and three books, including *Structural Foam* (Marcel Dekker, Inc.), he is a Senior Member of the Society of Plastics Engineers. Mr. Wendle received the B.S. degree (1960) in chemical engineering from the University of Idaho, Moscow.